ASHEVILLE-BUNCOMBE TECHNICAL INSTITUTE

NORTH CAROLINA
STATE BOARD OF EDUCATION
DEPT. OF COMMUNITY COLLEGES
LIBRARIES

D1768982

Discarded
Date JUN 25 2025

DESIGN OF MACHINE TOOLS

DESIGN OF MACHINE TOOLS

by OLAF A. JOHNSON

CHILTON BOOK COMPANY

PHILADELPHIA   NEW YORK   LONDON

Copyright © 1971 by
Olaf A. Johnson
First Edition
All rights reserved
Published in Philadelphia by Chilton Book Company
and simultaneously in Ontario, Canada,
by Thomas Nelson & Sons, Ltd.
ISBN: 0 8019 5508 4
Library of Congress Catalog Card Number 70-147253
Manufactured in the United States of America

Preface

This book is the result of a lifetime of practical experience directly connected with machine tool design. It is primarily intended for the young engineer entering the field but is also simple enough in presentation to be easily understood by the draftsman. Many valuable design problems are presented that will be helpful to the design engineer who is well advanced in the field.

In addition, this book will be valuable to persons in management, because to achieve the best results in designing machine tools, understanding and cooperation are necessary. The executive who must make the final decision as to whether or not a machine should be built will find material that will be helpful in appraising some of his problems.

Although much progress will continue to be made in automation, the sound principles of basic, fundamental machine design, which do not change, will continue to serve as its foundation.

I have endeavored here to cover those principles, to point out the pitfalls in design development and to deal thoroughly with details that contribute to good design. Some of these details may seem trivial to the inexperienced, and may even be overlooked by engineers with years of experience.

This book will help the designer to evaluate good design while the design is still in the preliminary stages. This will eliminate costly changes while the prototype machine is being built, or when mass production is started.

OLAF A. JOHNSON

Contents

Preface

Chapter 1 Steps in Design Procedure
- 1.1 Machine Tool Design, **1**
- 1.2 Market Survey, **1**
- 1.3 Preliminary Design, **1**
- 1.4 Prototype Design, **3**
- 1.5 Follow Up and Appraisal of Prototype Machine, **5**
- 1.6 Mass Production Machine, **5**

Chapter 2 Market Survey
- 2.1 Survey for Machine for Sale, **6**
- 2.2 Survey for Machine for your own Plant, **7**
- 2.3 Report covering Survey, **8**

Chapter 3 Preliminary Design
- 3.1 Outline of Machine, **12**
- 3.2 Evaluating Progress, **24**
- 3.3 Drive Diagram, **27**
- 3.4 Operating Sequence Diagram, **28**
- 3.5 Preliminary Layouts, **30**
- 3.6 Cost Analysis, **32**
- 3.7 Engineering Appraisal, **34**
- 3.8 Management Presentation, **36**
- 3.9 Machine Model, **36**
- 3.10 Reappraisal of Preliminary Design, **39**
- 3.11 Trials and Tests, **42**

Chapter 4 Prototype Machine
    4.1 Introduction, **44**
    4.2 Bearings in Machine Tool Design, **48**
    4.3 Mechanical Drive Components in Machine Tool Design, **77**
    4.4 Electrical Components in Machine Tool Design, **108**
    4.5 Pneumatics in Machine Tool Design, **125**
    4.6 Hydraulics in Machine Tool Design, **165**
    4.7 Material Selection and Heat Treatment in Machine Tool Design, **181**
    4.8 Transmission of Motion in Machine Tool Design, **189**
    4.9 Machining Practices for Machine Tool Design, **210**
    4.10 Good Fastening Practice and Good Proportioning of Fastened Parts in Machine Tool Design, **217**
    4.11 Main Spindle Layouts for Machine Tool Design, **229**
    4.12 Main Drive Assemblies for Machine Tool Design, **235**
    4.13 Main Layouts in Machine Tool Design, **242**
    4.14 Lubrication of Machine Tools, **243**
    4.15 Reappraisal of Prototype Machine, **248**
    4.16 Process Engineering for Machine Tools, **248**
    4.17 Detailing of Prototype Machine, **249**
    4.18 Release for Production, **250**
    4.19 Important Points to consider, **251**
    4.20 Final Appraisal, **251**

Chapter 5 Mass Production Machine
    5.1 Machine Improvements, **252**
    5.2 Assembly Drawings, **253**
    5.3 Detail Drawings, **253**
    5.4 Photographs, **254**
    5.5 Operating Instructions, **254**
    5.6 Floor Plans, **258**

Chapter 6 Reports and Communications
    6.1 Engineering Reports, **260**
    6.2 Engineering Communications, **262**

CHAPTER 1

Steps in Design Procedure

1.1 Machine Tool Design

Machine tool design is a challenging career full of changing problems; and because every project seems to have new problems, it is a career that requires the utmost in patience and perseverance. The right approach is a necessity in order to arrange and apply the various components for the best possible results. If one approach to the problem fails, another must be tried, and so on until the problem is solved. There is no room for doubt.

1.2 Market Survey

You may have been requested by management to design a new machine meeting certain requirements. If the machine is for sale, the sales department has probably made a survey to determine anticipated sales volume.

If the machine is for use in your own plant, the department requesting the new machine may have made a requirement survey.

If the idea for a new machine originated with you, make a market survey to determine the need for the machine.

MARKET SURVEY REPORT. When it is clear to you that a new machine would be advantageous to your company or client, a report should be prepared and presented to persons in authority for their consideration and decision.

1.3 Preliminary Design

The preliminary design can best be executed by the design engineer in charge of the project. It should be possible at this time for the designer to have in mind a clear idea of what the assemblies will be.

OUTLINE OF MACHINE. Make a number of freehand sketches representing different ways in which the machine can be built. It is always good practice to come up with your own ideas first, and then look into the design of other machines of a similar nature for the sake of comparison.

DRIVE DIAGRAM. You now have freehand outline sketches of the machine, and the market survey has probably revealed the speeds necessary to compete successfully with existing machines. The next step should, therefore, be a drive diagram. Make several freehand sketches, first showing pure mechanical drive, e.g., gears, cams, shafts, levers, push and pull rods, ratchet drives, belt drives, chain drives. The drive is the important part of the machine and enough time should be devoted to this to make sure you have a good drive. Also investigate whether it is possible to supplement part of the mechanical drive with hydraulic, pneumatic or electrical components.

OPERATING SEQUENCE DIAGRAM. As soon as you have enough information in sketches and notes, draw an operating sequence diagram. This will show when all the functions of the machine take place in relation to each other.

PRELIMINARY LAYOUTS. Make full size sectional layouts of the main spindles and important parts of the drive. Just enough detail should be worked out to make sure there is enough room for the complete drive. Then, full size outline layouts should be made of all four sides and the plan view of the machine. If desired, a perspective view of the machine may also be made at this time.

COST ANALYSIS. You now have gathered enough information to estimate how much it will cost to build a prototype machine.

ENGINEERING APPRAISAL. A complete report should be written, which is comprehensive enough to include all the facts, and covers the main construction of the machine, outstanding features, operation, performance and cost. The appearance of the machine is covered by full size outline drawings and perspective view. Now you are well prepared to discuss the machine in detail with your engineering superior and associates and to consider their criticisms. Have an open mind, and reevaluate your design thoroughly. Major changes are easy to make at this stage, but do not make any changes unless you see definite advantages.

1.4 Prototype Design

MANAGEMENT PRESENTATION. The next step is to revise your preliminary layouts and your written report if necessary, for presentation to management. Perhaps, at this time, management will be impressed, and would like to see a model of the machine.

MACHINE MODEL. A machine model can easily be made at little expense. This model may be made to a reduced scale; but it is often found more desirable to have a full size model for better visualization.

If the machine is for sale, management and the sales department now have their opportunity to make suggestions for better, more attractive appearance.

If the machine is for use in your own plant, the department requesting the machine has an opportunity to offer suggestions.

The manufacturing departments now also should be invited to view the model and may have many suggestions that would influence the building and cost of the machine.

REAPPRAISAL OF PRELIMINARY DESIGN. After your preliminary layouts and machine model have been revised to satisfy all concerned, your report should also be revised, if necessary. Now a second presentation to your superior and associates in the engineering department is necessary. Management and the sales and manufacturing departments should also be informed of the latest developments, so that if decided on favorably, permission can be granted to go ahead with the prototype design.

1.4 Prototype Design

The best procedure is to have several designers or engineers working as a team on the various assemblies of the machine. Some of the assemblies overlap, and are dependent on each other.

LUBRICATION. It is important to keep good lubrication in mind all through the design work. Faulty lubrication can cause the failure of any design.

MAIN SPINDLES. The main spindles or important shafts should now be laid out whenever possible in full size sectional views. You have preliminary layouts of these views, and should now consider every detail when making the final layouts.

The forging manufacturers should be contacted early to make sure they can meet your manufacturing schedule.

MAIN DRIVE ASSEMBLIES. Considerable time and care should be

devoted to these assemblies. You have worked out a simple drive with your freehand sketches, but you must now consider problems relating to practical, economical manufacturing, reasonably easy assembly procedures, and proper lubrication.

MAIN LAYOUTS. You may find it to great advantage to have a few main layouts which would take in more than one assembly, and may even cover the entire machine. They would give you a chance to investigate thoroughly whether all parts can be readily assembled.

REAPPRAISAL OF PROTOTYPE DESIGN. As the design has progressed, you may have had to make some changes in appearance of the machine. You may also have made changes in the drive of the machine and the operating sequence. The model should, therefore, be brought up to date, and final layouts should be made of the drive diagram and the operating sequence diagram. The written report covering operation and functional points on the machine should also be brought up to date.

The cost analyst may now be interested in reappraisal of his cost estimates, because at this time the cost of purchased parts or components may be available, and a better, clearer picture is available of the whole machine.

PROCESS ENGINEERING DEPARTMENT. The process engineering department should be contacted as soon as you have layouts of some of the parts that may require special handling or procedures.

DETAIL DEPARTMENT. Close cooperation is also required between the detail department and the engineer or designer in charge of the project. The designer, in the process of designing, has come to definite decisions regarding material, methods of manufacturing, heat treatment and assembly methods, something that could not be decided just by glancing over the layouts. Therefore, the detailer should closely follow the layouts, and if in doubt, should consult the designer.

RELEASE FOR PRODUCTION. A little time spent in checking the detail drawings will pay off. Remaking of parts, because the detail drawings are not clear or are inaccurate or faulty, is costly and causes unnecessary delay and confusion.

Close contact with the purchasing department is strongly recommended, and as soon as you have enough information on purchased components, the purchasing department should be asked to get prices and delivery dates. Then all drawings requiring patterns for cast parts should be released.

1.4 Prototype Design

1.5 Follow Up and Appraisal of Prototype Machine

It is vitally important to the success of a new machine that the engineer or designer checks up on the building of the prototype machine.

APPRAISAL OF THE PROTOTYPE MACHINE. A complete written report is in order by the follow up man or the design engineer covering all changes of any importance, performance of the machine and suggested plans for future design, if changes had not been taken care of on the prototype machine. This report is important for your active file. A concise report may be formulated from this and presented to your superior and management.

1.6 Mass Production Machine

Quite often, the mass production machine is turned over to another engineer, thus relieving the original designer for further creative work.

MODEL OF THE MASS PRODUCTION MACHINE. The model for the prototype machine should be kept until the mass production machine has been released and has proved successful.

CHAPTER 2

Market Survey

2.1 Survey for Machine for Sale

The need for a new machine may come to the attention of the designer in various ways:

1. In the course of conversation with a salesman, machine demonstrator, or some member of his own organization, he may suddenly see where a new machine can be put to good use. The other parties may not have the slightest thought in that direction, but the designer's mind recognizes the opportunity. He makes notes and freehand sketches while the idea is fresh and files them for future reference.
2. Management may have been requested to develop a new design. An alert salesman may have discovered that the machines a customer is using are old and could be greatly improved to the advantage of the customer.
3. The request for a new machine may, of course, come from a company that has developed a new part, and they therefore need a new machine to make it. Since manufacturing the part may involve cutting, grinding, forming or other die work, and assembling, more than one machine would, of course, be required.

In any event, the exact nature and purpose of the machine must be determined. Such factors as manufacturing speeds, composition of material to be worked, and others must be taken into account. All this requires that a thorough survey be made of the job to be done and the conditions under which it must be done.

2.1 Survey for Machine for Sale

2.2 Survey for Machine for your Own Plant

The request for this machine would come, most likely, directly from the department in need of the machine. The size and type of your organization would determine how the machine should be designed and processed.

1. If your company is in the machine tool industry, your engineering department would have qualified engineers to undertake the project and also have manufacturing departments properly equipped and of adequate size to take on the extra work. This is an ideal setup, because your design engineers can have close contact with the manufacturing departments, and with the department requesting the machine.

    Under this setup, it is important to have a department where the machine can be thoroughly tested before being delivered to the department requesting the machine.

Figure 2.3–1. A group of Gleason generators used in rough and finish machining of Oldsmobile differential drive pinion. (Courtesy of General Motors Corporation)

Market Survey

Figure 2.3-2. Performing the finish turn operation on the Oldsmobile ring gear blank. From this point, conveyor carries the gear to the broach machine. (Courtesy of General Motors Corporation)

2. It often pays to engage a reputable consulting engineer to guide your engineering department along with his practical knowledge, or to have him undertake the complete design. The machine could then be built in your own shop if adequately equipped, or in an outside machine shop.

2.3 Report Covering Survey

It is important to keep all notes and sketches even though some have been superseded. Keep them and mark them superseded. Using this material, make one detailed report for your own file. This should cover all details. You may have been to a plant employing automation and have seen a group of machines laid out so that a conveyor system easily takes care of the flow of parts from one group of machines to another. It is up to you, in designing a new machine, to make notes and freehand sketches of such a layout, so that the machine you are

2.3 Report Covering Survey

designing will adequately meet all requirements and fit into the layout. (Figures 2.3-1 to 2.3-4)

From this detailed report, write a concise report covering just the most important points. This report goes to management and your immediate superior. They are, at this time, vitally interested in comparing production rates and cost with those of present machines or methods.

If the machine is for your own plant, the production rate must be higher, accessibility to replaceable parts must be better, quality of parts manufactured must be better, or, if the new machine will replace several other machines, the advantages of such replacement must be obvious.

Figure 2.3-3. Group of machines for drilling and tapping Oldsmobile differential ring gears. (Courtesy of General Motors Corporation)

Market Survey

As a whole, these qualifications may also apply to machines for sale, and should be covered in the report. Remember, that you as a designer will, as far as reasonable, be held responsible for the statements or estimates that you make at this time. Do not come up with figures or promises that will look good to management, but be so far out of line that it would be impossible to meet them in your proposed design.

Figure 2.3–4. Group of Oldsmobile cylinder block line machines, for spot facing and roughing the cam bores, and spot facing the crankshaft bore. (Courtesy of General Motors Corporation)

2.3 Report Covering Survey

CHAPTER 3

Preliminary Design

3.1 Outline of Machine

It happens often to the best of designers that they are so engrossed in putting across a new idea or a new method of manufacturing that the principles of basic, fundamental machine design are overlooked. When the new idea or new method of manufacturing fails, it is likely to be due to some minor oversight. It is, therefore, always wise to spend time analyzing the basic design. Always remember that your management or your client is interested in reasonably fast results, and this can only be accomplished if the design has been thoroughly analyzed before being released from the engineering department for production.

You now have an order to go ahead with the preliminary design. This work is primarily done to have something tangible to grasp at. The market survey has revealed exactly what the new machine is intended to do. You are probably in possession of some sample parts, a few parts covering the range, and detail drawings showing you the full range. You may also have specifications giving accuracy required, kind of material, and the desired speed at which the operation should be performed. If the machine is to be used in an automation setup, where several groups of machines are involved, you would also have the size of the machine within a certain maximum, e.g., maximum floor space and maximum height. (Figure 3.1–1) In designing machine tools, you may be requested to design a machine or piece of equipment

Figure 3.1–1a. Drilling and honing of engine ports. (Courtesy of Ford Motor Company)

3.1 Outline of Machine

to replace or to be added to existing equipment as shown in Figures 3.1–1a, 3.1–1b, 3.1–1c or 3.1–1d. Also see Figures 2.3–1 to 2.3–4.

To have something tangible to start with, we will assume that we have approval to conduct a preliminary investigation for the design of a machine to cut the tooth spaces in a straight bevel gear. We have

Figure 3.1–1b. Crankshaft balancing. (Courtesy of Ford Motor Company)

Figure 3.1–1c. Machining operations on Oldsmobile differential carrier. (Courtesy of General Motors Corporation)

Figure 3.1–1d. Machining operations on Oldsmobile differential carrier. (Courtesy of General Motors Corporation)

3.1 Outline of Machine

Preliminary Design

sample gears and drawings showing all limitations. (Figures 3.1–2 and 3.1–3)

We will also assume that one tooth space will be finished in one turn of the cutter. The next thing to be determined would be the diameter, shape and mounting of the cutter. In a large machine tool

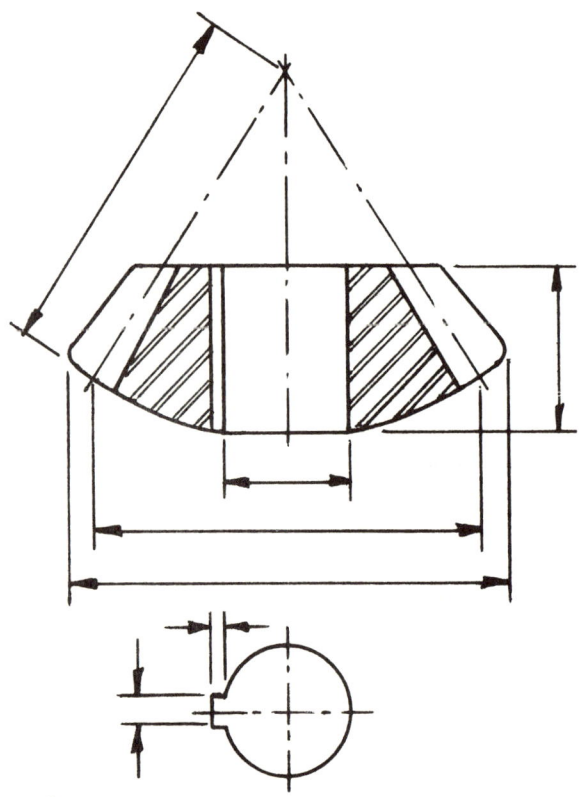

Figure 3.1–2. Typical sample pinion and minimum requirements. This drawing shows a pinion, the smallest gear in a pair. The dimensions and information required are typical.

STRAIGHT BEVEL TEETH

No. of Teeth _____	Addendum _____	Chord. Addendum _____
Pitch (D.P.) _____	Whole Depth _____	Chord. Thickness _____
Pitch Diameter _____	Cone Distance _____	Speed—RPM _____
Pitch Angle _____	Face Angle _____	Backlash in Asb. _____
Shaft Angle _____	Root Angle _____	Mate—Dwg. No. _____
Pressure Angle _____	Tooth Angle _____	

3.1 Outline of Machine

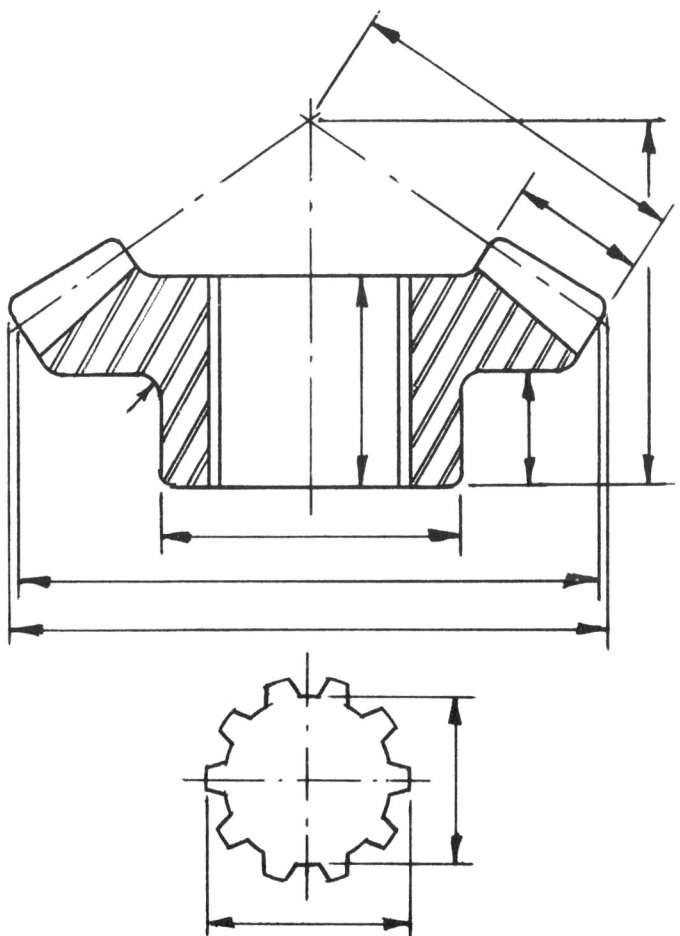

Figure 3.1-3. Typical sample gear and maximum requirements. This drawing shows a gear, the largest member of a pair. The dimensions and information required are typical.

STRAIGHT BEVEL TEETH

No. of Teeth _____   Addendum _____   Chord. Addendum _____
Pitch (D.P.) _____   Whole Depth _____   Chord. Thickness _____
Pitch Diameter _____   Cone Distance _____   Speed—RPM _____
Pitch Angle _____   Face Angle _____   Backlash in Asb. _____
Shaft Angle _____   Root Angle _____   Mate—Dwg. No. _____
Pressure Angle _____   Tooth Angle _____

Preliminary Design

company, it is likely that this would not be determined by the machine designer, but by other specialists. There may be several cutter sizes required to cover the entire range of the machine. (Figure 3.1–4)

The size of the cutter would be determined by the amount of material to be removed within a certain time limit, the toughness of the material to be removed, and the life expectancy of the cutter.

Knowing the size limitations of the gears and cutters, you are now in a position to do some exploratory work. First make a couple of full size layouts, one showing the cutter and gear limitations for the small gear, e.g., the pinion, and one showing the cutter and gear limitations for the mating gear. (Figures 3.1–5 and 3.1–6) These layouts

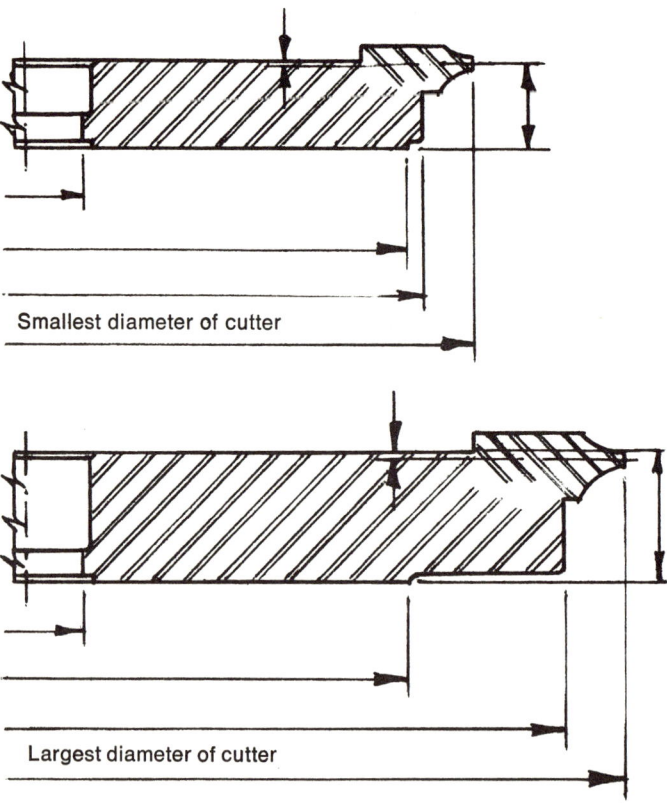

Smallest diameter of cutter

Largest diameter of cutter

Figure 3.1–4. Anticipated range of cutter sizes. These drawings should reveal maximum and minimum diameters, which are determined from type and amount of material removed per tooth space, and type of material used for the cutter blades.

3.1 Outline of Machine

should be accurate, but need not be elaborate. A few simple lines at this time will furnish sufficient information. You can now readily see from these layouts the direction of all adjustments and motions. While

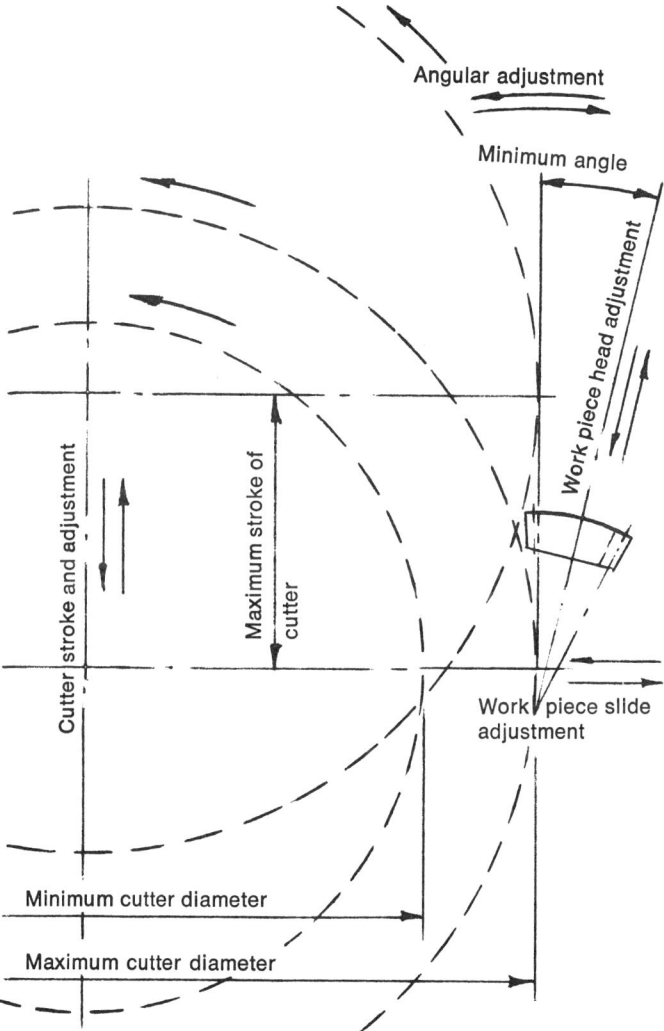

Figure 3.1–5. Minimum root angle adjustment and other important adjustments and motions. Mark down at this time the machine member you think should be adjusted to meet the requirements. As you progress with your sketches, you may, however, find it more advantageous to adjust other members to meet the same requirements.

Preliminary Design

you are making layouts, you can be thinking about the operation of the machine, and as thoughts come to you, make notes and sketches. By the time the layouts are finished, you have probably come up with

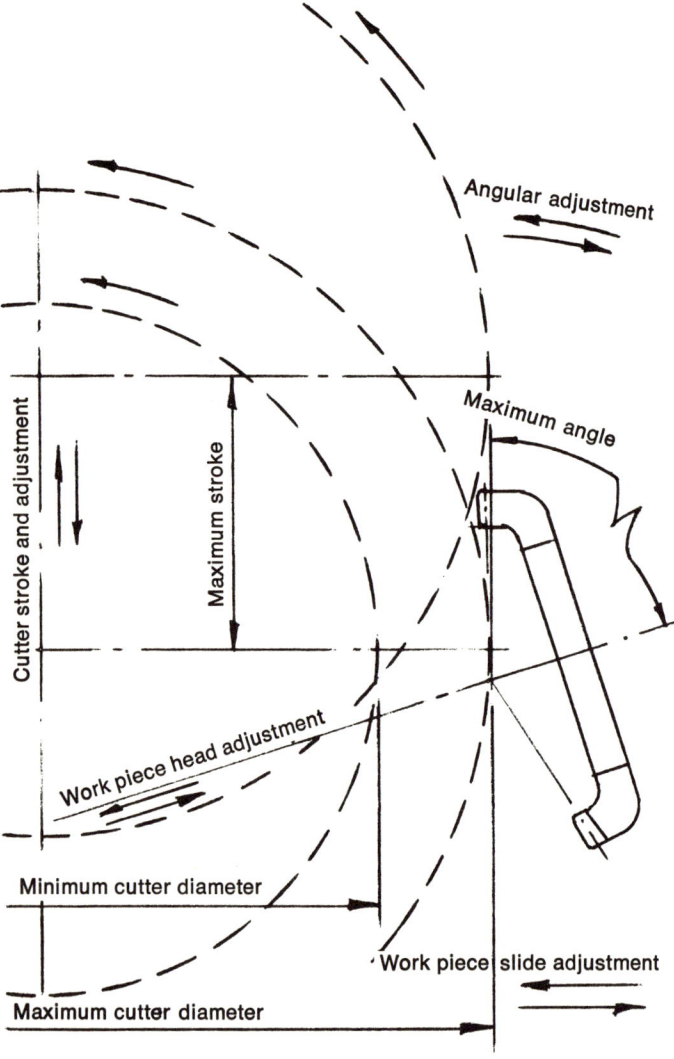

Figure 3.1–6. Maximum root angle adjustment and other important adjustments and motions. Mark down at this time the machine members you think should be adjusted to meet the requirements. As you progress with your sketches, you may, however, find it more advantageous to adjust other members to meet the same requirements.

3.1 Outline of Machine

some definite ideas of the appearance of the machine, and probably also have a rough idea of how many machine assemblies the machine should be divided into. Although this is not absolutely necessary at this time, it is good to start thinking about it early. The essential thing now is to be sure that you have considered all motions and adjustments required.

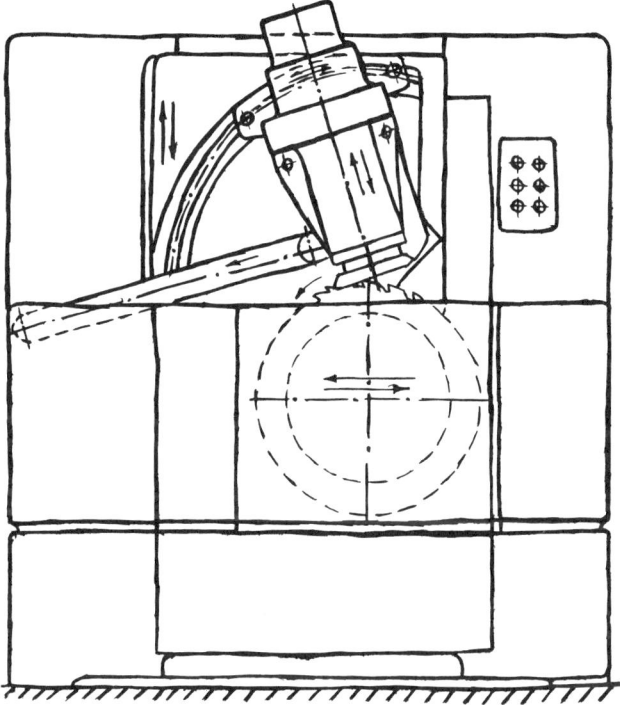

Figure 3.1-7. Freehand sketch of proposed machine. This freehand sketch may be the result of several rough freehand sketches.

1. The loading and unloading of the work piece could in this case be easily performed by hand or automation, since the parts are assumed to be light in weight.

2. The cutter is easily replaced by opening or removing a door.

3. The cutting oil, or in machine tool parlance, the coolant (in some cases a water soluble solution is used) may easily be applied at the cutting edges and returned to the frame of the machine, which with proper screening, may be used for the coolant.

4. The chips may here be conveyed to the back of the machine, while the coolant is removed from the chips.

5. As indicated by the arrow, the cutting forces may be absorbed in the frame of the machine through an adjustable member located on the center of the work piece, in line with the cutting forces.

Preliminary Design

Table 3.1–8. Average Unit Power Requirements *

HORSEPOWER PER CUBIC INCH PER MINUTE

MATERIAL	Brinell hardness number or R_c as indicated	UNIT POWER † Horsepower per Cubic Inch per Minute		
		TURNING P_t High speed stl. and Carbide tools Feed .008 to .010 inch per revolution	DRILLING P_d High speed steel drills Feed .002 to .010 inch per revolution	MILLING P_m High speed stl. and Carbide tools Feed .004 to .008 inch per tooth
Steels				
Wrought & Cast	85–200	1.2	1.0	1.2
Plain Carbon stl.	35–40 R_c	1.6	1.4	1.5
Alloy steels	40–50 R_c	2.0	1.7	2.0
Tool steels	50–55 R_c	2.2	2.0	2.2
Cast Irons				
Gray, Ductile	110–190	1.0	1.0	1.0
and Malleable	190–320	1.8	1.6	2.0
Stainless steels Wrought & Cast Ferritic	135–275	1.5	1.3	1.4
Austenitic and Martensitic	30–45 R_c	1.5	1.4	1.6
Precipitation hardening Stainless steels	170–450	1.5	1.4	1.7
Titanium	250–375	1.0	1.0	1.2
High Temperature Alloys—Nickel and Cobalt Base	200–360	2.0	2.0	2.5

You can now make several freehand sketches, outlining the machine. The cutting motions and adjustments will in this case determine the limitations in one view. You now have to decide whether to make this view the plan view or side view, or to arrange it in an angular position. Make freehand sketches of each condition and make a thor-

3.1 Outline of Machine

MATERIAL	Brinell hardness number or R_c as indicated	UNIT POWER † Horsepower per Cubic Inch per Minute		
		TURNING P_t High speed stl. and Carbide tools Feed .008 to .010 inch per revolution	DRILLING P_d High speed steel drills Feed .002 to .010 inch per revolution	MILLING P_m High speed stl. and Carbide tools Feed .004 to .008 inch per tooth
Refractory Alloys				
Tungsten	321	3.5	3.3	3.6
Molybdenum	229	2.4	2.0	2.5
Columbium	217	2.2	1.7	2.3
Tantalum	210	2.7	2.1	2.7
Nickel Alloys	80–360	2.2	2.2	2.6
Aluminum Alloys	30–150 500 kg.	0.3	0.2	0.4
Magnesium Alloys	40–90 500 kg.	0.2	0.2	0.2
Copper	80 R_b	1.2	1.1	1.2
Copper Alloys	20–80 R_b 80–100 R_b	0.8 1.2	0.6 1.0	0.8 1.2

* From "Machining Data Handbook," Metcut Research Associates, Inc., Cincinnati, Ohio.
† Power requirements at spindle drive motor, corrected for dull cutter and 80% spindle efficiency.

ough study of each case. Disregarding the cutting forces at this time, you should consider the following:

1. Loading and unloading of work parts
2. Replacement of cutter
3. Application of coolant
4. Accumulation and removal of chips

There may also be other minor considerations. When you have

Table 3.1-9. Horsepower Required at Motor *

	Turning Operations	Milling Operations	Drilling Operations
Cutting Speed, feet per min.	$V_c = .262 \times D_t \times \text{rpm}$	$V_c = .262 \times D_m \times \text{rpm}$	$V_c = .262 \times D_d \times \text{rpm}$
Revolutions per minute	$\text{rpm} = 3.82 \times \dfrac{V_c}{D_t}$	$\text{rpm} = 3.82 \times \dfrac{V_c}{D_m}$	$\text{rpm} = 3.82 \times \dfrac{V_c}{D_d}$
Feed Rate, inches per min.	$f_m = f_r \times \text{rpm}$	$f_m = f_t \times T \times \text{rpm}$	$f_m = f_r \times \text{rpm}$
Feed Per Tooth, inches		$f_t = \dfrac{f_m}{T \times \text{rpm}}$	
Cutting Time, minutes	$t = \dfrac{L}{f_m}$	$t = \dfrac{L}{f_m}$	$t = \dfrac{L}{f_m}$
Rate of Metal Removal, cubic inches per min.	$Q = 12 \times d \times f_r \times V_c$	$Q = w \times d \times f_m$	$Q = \dfrac{\pi D_d^2}{4} \times f_m$
Horsepower Required at Motor	$HP_s = Q \times P_t$	$HP_s = Q \times P_m$	$HP_s = Q \times P_d$

3.2 Evaluating Process

chosen the sketch that is the most favorable to all conditions considered, make a freehand sketch to determine the appearance of the machine. At this time you have to consider cutting forces to be sure that they can be adequately absorbed. (Figures 3.1–7 to 3.1–9)

3.2 Evaluating Progress

You have now made a study of the machine, and in addition to freehand sketches, have made notes of your conclusions. At this time it may be good to find out what has been done by someone else for a machine of similar nature. By comparison, you can readily evaluate your own design.

Also disclose your plans to associates and superiors. Their reactions can influence your future course. If they make suggestions that will improve the design, now is the time to make these changes. Quite often, criticism is offered with no suggestions for improvement. There may be conditions, seemingly unimportant ones, that you have overlooked. It is always easy for someone else to find faults, but have an open mind for all criticism. Other people may not have the overall picture that you have at this time. However, some conditions that could be potential trouble may be obvious to them. Therefore, consider all criticism. If given with no offer for remedy, you have food for

---

* From "Machining Data Handbook," Metcut Research Associates, Inc., Cincinnati, Ohio.

SYMBOLS

D_t = Diameter of work piece in turning—inches
D_m = Diameter of milling cutter—inches
D_d = Diameter of drill—inches
d = Depth of cut—inches
f_m = Feed rate—inches per minute
f_r = Feed—inches per revolution
f_t = Feed—inches per tooth
L = Length of cut—inches
P_t = Unit power required in turning—Horsepower per cubic inch per minute. Corrected for dull cutter and 80% spindle drive efficiency
P_m = Unit power required in milling—horsepower per cubic inch per minute. Corrected for dull cutter and 80% spindle drive efficiency
P_d = Unit power required in drilling—horsepower per cubic inch per minute. Corrected for dull cutter and 80% spindle drive efficiency
Q = Rate of metal removed—cubic inches per minute
rpm = Revolutions per minute of work or cutter
t = Cutting time—minutes
T = Number of teeth in cutter
V_c = Cutting speed—feet per minute
w = Width of cut—inches
HP_s = Horsepower required at motor

Preliminary Design

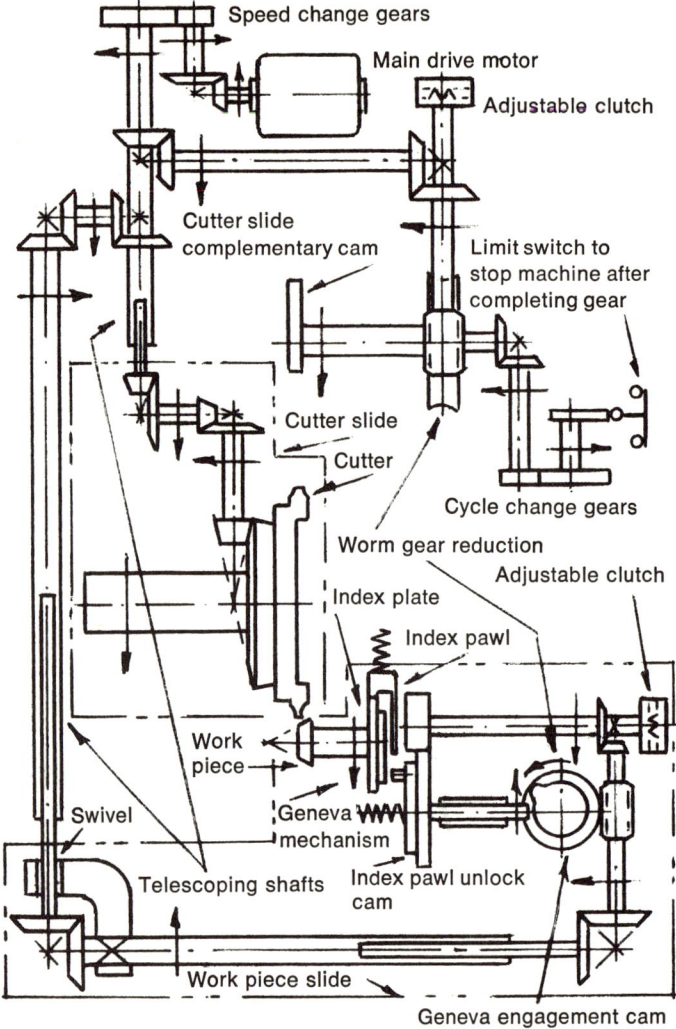

Figure 3.3-1. Drive diagram. The drawing of this drive diagram is not to any exact scale. Several freehand sketches, not shown, have led up to this drawing. It is a necessary and handy reference in the process of designing the various parts of the machine.

From the maximum cutting forces, Figure 3.1-8, you have determined the horsepower requirements of the drive motor, and from the market survey you have determined the speed of the cutter, which is mainly governed by the material being cut, the material used for the cutter blades and the economical number of tooth slots being cut between cutter sharpenings.

You can, therefore, determine the minimum diametral pitch of the gears in

3.3 Drive Diagram

thought. You may not have a good solution immediately, but if you have not closed your mind, it will come, as you are working out other problems.

3.3 Drive Diagram

Your survey has given you the speeds desired, and the preliminary outline drawings of the machine show you all the required adjustments and motions which are necessary for the successful operation of the machine. With this information, you are in a position to start preliminary sketches of the drive. (Figure 3.3–1) The drive is basic to your machine, so even though you are preparing only a preliminary design at this time, you should give it enough thought to be reasonably sure it can be worked out successfully.

We will assume that the best final drive for the cutter would be a pair of bevel gears. The type of bevel gears does not have to be determined at this time. From the size of cutter diameters, you can determine the size of the large gear, and the diameter of this gear should, of course, be as large as practical. The cutting forces are dependent on the type of material to be cut, the amount to be removed and the speed of removal. This will determine the pitch of the gear teeth and, therefore, also the number of teeth required.

The work piece, in this case, should be held securely while cutting, and should be indexed accurately from tooth space to tooth space. There are several ways of doing this. We will assume that the work piece is held securely to the work piece spindle with a hydraulic or pneumatic chuck, that the work piece spindle is kept from rotating while cutting, and that there is an accurately finished index plate. We will also assume that the indexing from one slot to the next in the index plate is performed with a geneva drive.

the drive from the motor to the cutter, and also to the cutter slide cam. The diametral pitch of the gears indexing the work piece is determined by the speed and the• mass being moved. The final drive gear of the cutter should be as close to the cutter diameter as the design will permit, and cannot be determined definitely until the full size prototype layouts are made.

A successful completion of the design may also make it necessary to change a pair of gears which then would reverse the drive direction. The simplest solution then, to keep the desired direction, would be to reverse another pair of gears in the train. If this is impossible or undesirable, an idler gear may be used to reverse the direction of the drive.

Preliminary Design

From the maximum cutting forces you can now determine the size of the drive motor, and we will assume that the rpm of the motor is 1800. (Figure 3.1-9) The cutting forces and speeds will, of course, determine the size of the drive shafts. With antifriction bearings, you can run the drive shafts relatively fast to keep the size of the diameters small. There are other considerations determining the size of the shafts such as length of shafts. For the preliminary sketches of the drive, it is not necessary to consider these points.

3.4 Operating Sequence Diagram

This diagram is worked out from the calculations you have already made regarding the various functions of the machine. They may include mechanical functions, hydraulic functions, pneumatic functions and the like. It will show for one cycle of the machine, the instantaneous position of the work piece in relation to the cutter, or a number of cams or parts of the index mechanism in relation to other parts of the drive or, the relationship to other functions. It will cover a number of functions, and will enable even the inexperienced to digest, in a short time, the entire operation of the machine.

A lengthy report, or lengthy operating instructions, which are necessary in the process of designing a complicated machine, are diffi-

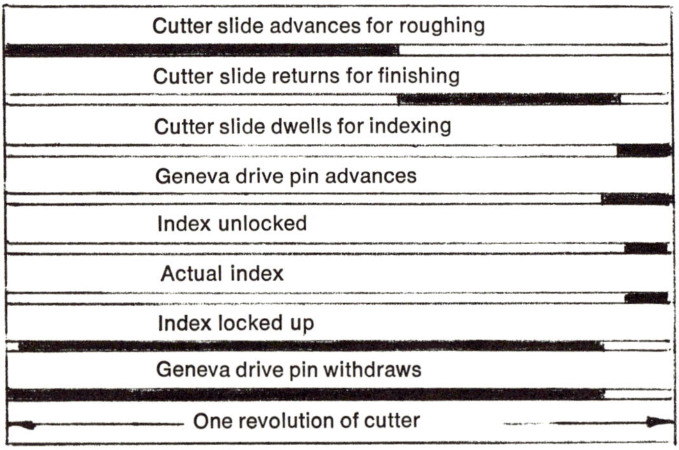

Cutter slide advances for roughing
Cutter slide returns for finishing
Cutter slide dwells for indexing
Geneva drive pin advances
Index unlocked
Actual index
Index locked up
Geneva drive pin withdraws
One revolution of cutter

Figure 3.4-1. Operating sequence diagram. This diagram is based on time, and is considered much easier to read than a circular diagram based on angular motion. It shows one revolution of the cutter, which in this case is the basic cycle of the machine. All functions are indicated in black.

3.4 Operating Sequence Diagram

cult for the average person to follow. Therefore, the little time spent on an operating sequence diagram (Figure 3.4–1) is well paid for in making the presentation of the machine so much easier to understand.

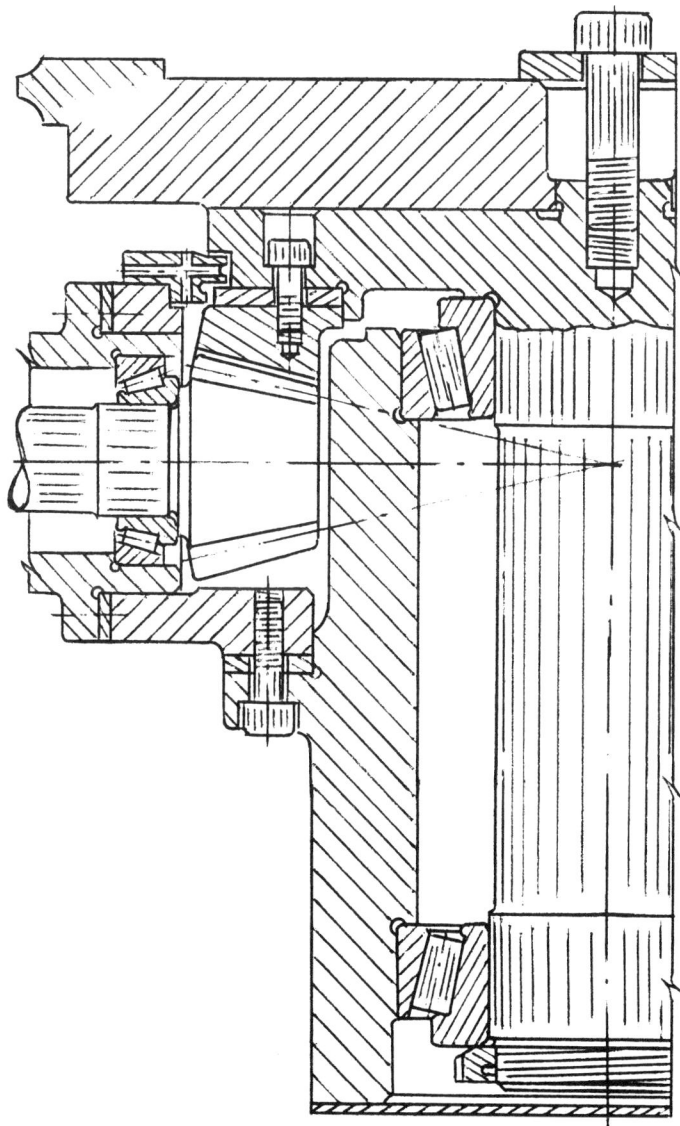

Figure 3.5–1. Cutter spindle preliminary layout. Just enough should be shown on this layout to make sure the design is practical.

Preliminary Design

3.5 Preliminary Layouts

Now that you have accumulated enough information in notes and sketches, you should be ready to make full size preliminary layouts.

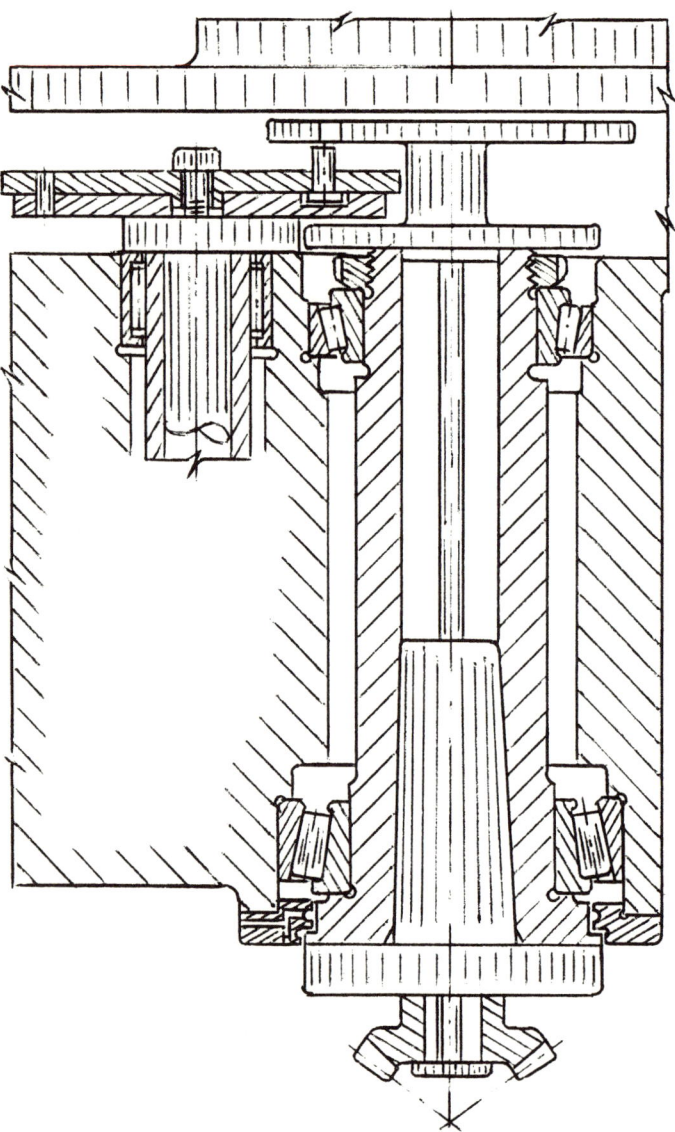

Figure 3.5–2. Workpiece spindle preliminary layout. Just enough of the layout has been completed to be sure that the design is practical.

3.5 Preliminary Layouts

The important layouts in the case at hand would be the cutter spindle and the work piece spindle.

CUTTER SPINDLE. The speed and the cutting forces (See figure 3.1–8) can be calculated from the information you already have. There are many choices of bearings. In some cases, one type is just as good as the other type. In other cases, just one type is best, considering all conditions.

There is not much difference in the space occupied by each type of bearing, so in this case, we will assume that tapered roller bearings

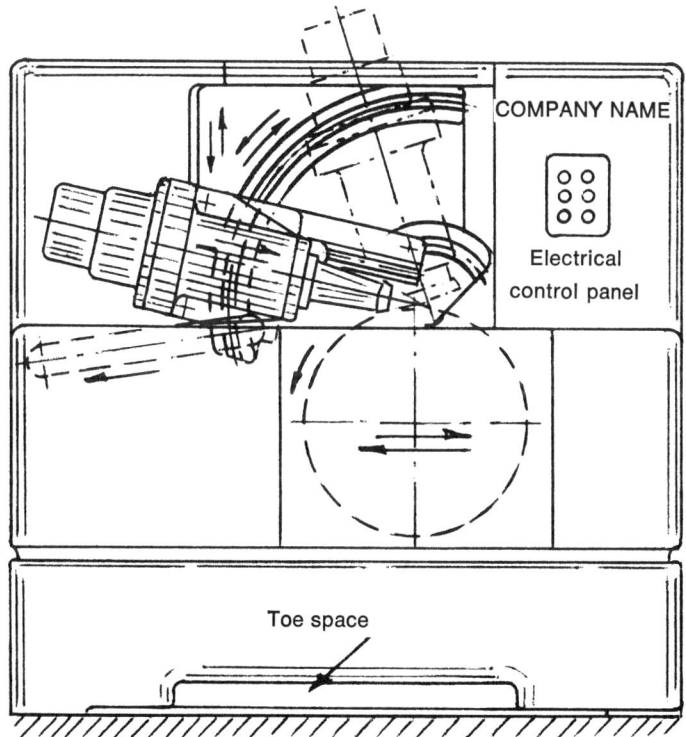

Figure 3.5–3. Front view preliminary layout of proposed machine. The arrows show:
Vertical adjustment of work piece slide
Angular adjustment of work piece head
Cone distance adjustment of work piece head
Horizontal travel and adjustment of cutter
Direction of cutter rotation
How cutting forces are directed to be absorbed in the frame of the machine through an adjustable support rod

Preliminary Design

have been selected. Not too much time should be spent in the selection of bearings at this time, because a change can easily be made when you come to the prototype design. At any rate, let us assume, considering manufacturing, assembling, speeds, loads, and cost, that this is the wise selection. (Figure 3.5–1)

WORK PIECE SPINDLE. The cutting forces (Figure 3.1–8) are also here and the main loads are to be considered under static conditions. We will in this case also assume the use of tapered roller bearings just to get a preliminary outline of the spindle housing. (Figure 3.5–2)

You are now ready to make full size preliminary layouts of the plan view of the machine and all four sides, and if deemed necessary, a perspective view to a reduced scale. The full size layouts should show the outline of the machine in heavy lines, and the outline of doors and compartments, such as that for the electrical compartment. (Figures 3.5–3 to 3.5–8)

3.6 Cost Analysis

The information gathered so far in full size preliminary layouts, sketches, and notes is enough for an experienced cost analyst to come up with a fairly accurate cost of the finished machine. From your drive diagram he can get the number of gear pairs required, cams, other proposed mechanisms, and size of motor and electrical controls. In many cases, the designer, himself, has been in close enough contact

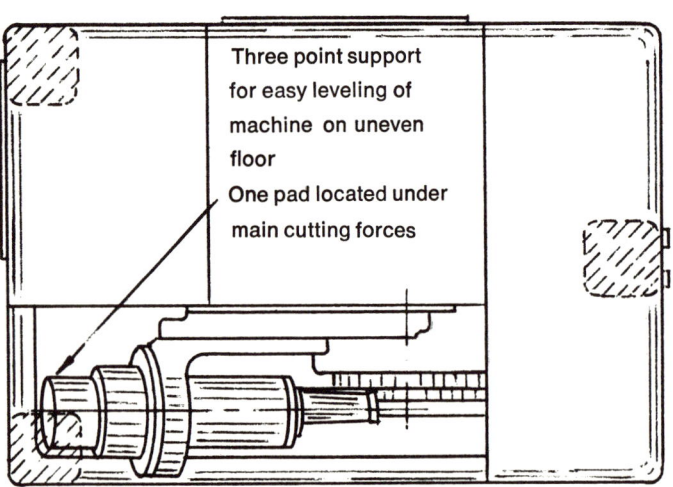

Figure 3.5–4. Plan view preliminary layout of proposed machine.

3.6 Cost Analysis

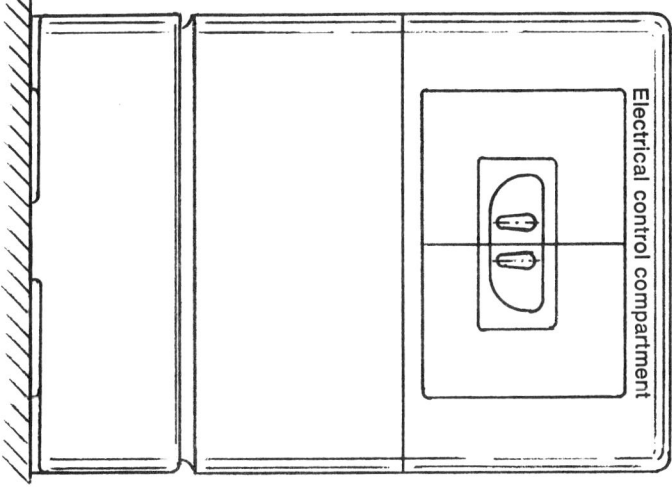

Figure 3.5-5. Right side preliminary layout of proposed machine.

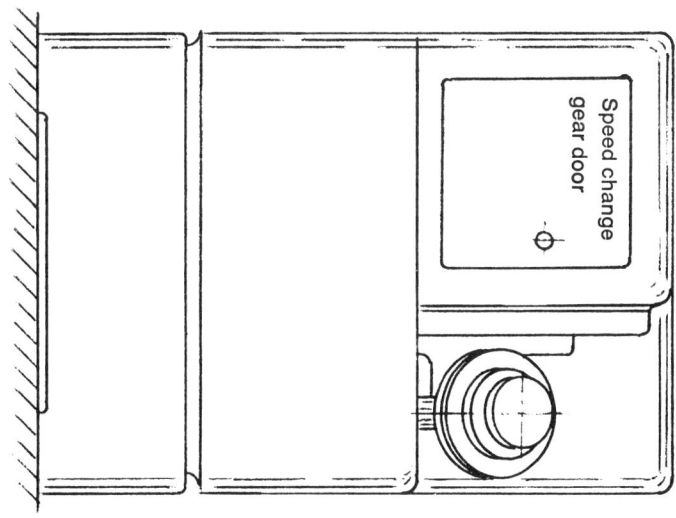

Figure 3.5-6. Left side preliminary layout of proposed machine.

Preliminary Design

with the manufacturing departments and is familiar enough with the prevailing cost per hour charged for the use of the various manufacturing machines to arrive at a reasonably close cost figure. At this time, you probably also have preliminary cost figures on most of the purchased parts.

3.7 Engineering Appraisal

It is now a good time to reappraise your design. Analyze all gears and decide, considering speed, noise, accuracy requirements and any other factors, which type you propose to use in your machine.

If you have gear specialists in your own organization, the wisdom of your choice can easily, and quickly be confirmed. If, on the other hand, you have to consult specialists by mail or telephone, this should be done at the time you are sketching your drive diagram.

You have now made a thorough study of your machine, and you have accumulated much more in your mind than you have had a

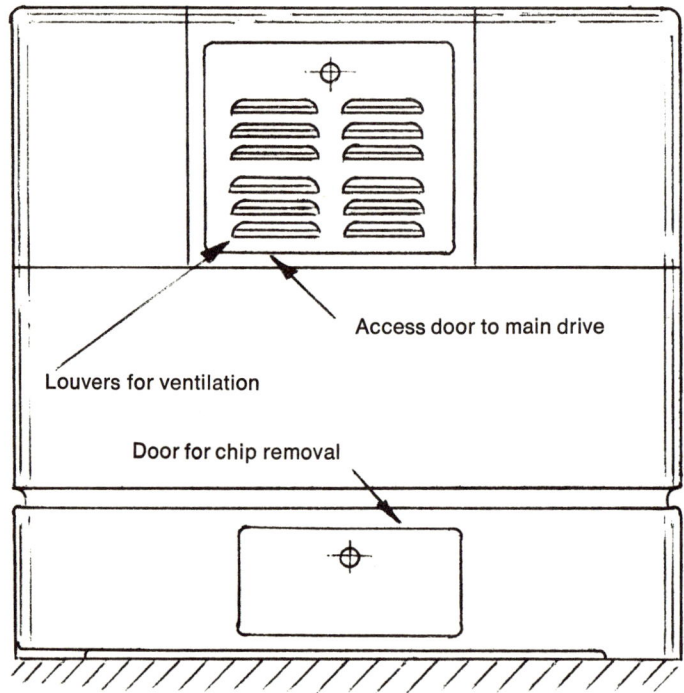

Figure 3.5–7. Rear view preliminary layout of proposed machine.

3.7 Engineering Appraisal

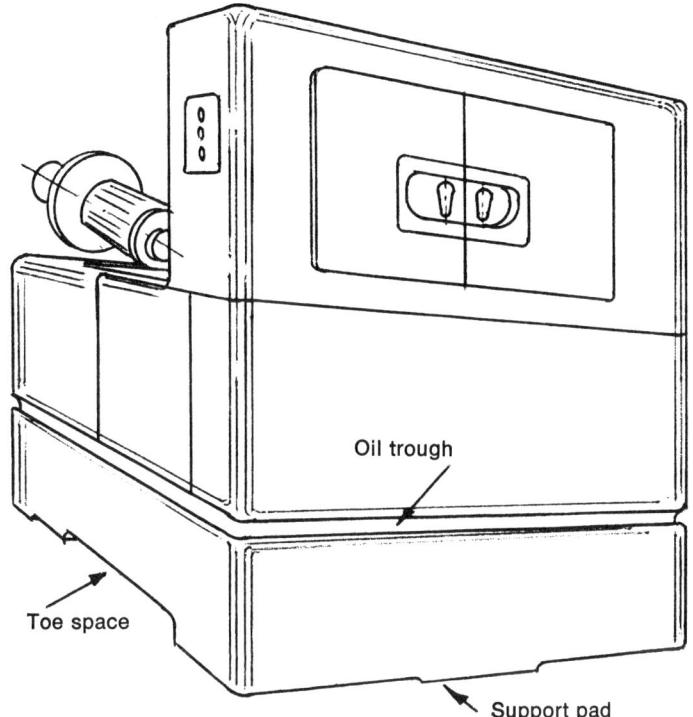

Figure 3.5–8. Perspective view of proposed machine.

chance to get down on paper. Many problems have come to your attention such as manufacturing, assembling, heat treatment and cast parts pattern making. It is now a good time to get the experts in these departments acquainted with your design. If the machine is to be used in your own plant, it would also be wise to invite someone in authority from the department where the machine is to be used to get acquainted with your design. As a wise designer, you have, of course, also talked to the operator of the present equipment, to enable you to evaluate your design.

If the men you have dealt with have suggestions to make, and you agree that these suggestions will be profitable, you now have a chance to make alterations, before presenting your design to your superiors or management.

If you and the specialists do not agree on some point, and you feel that you have a good reason to go ahead as planned, do not ignore their arguments, but take it up with your associates, superiors and, if desirable, management. Otherwise, you will have a troublesome situation to deal with even though you may be entirely correct in your decision. Make a complete written report at this time to cover all the important facts about the machine, and why you consider your design the best arrangement. Include outstanding features of the design, a description of the machine's operation, the expected performance, and the cost to build the prototype machine.

3.8 Management Presentation

After your written report has been evaluated by your engineering department and you have made desirable changes, write a concise report that will cover all the important functions of the machine, the quality expected of the parts to be produced, and the estimated cost of the machine. The layouts will cover the appearance of the machine. In addition, a complete specification sheet should be prepared, covering size and capacity of the machine and its auxiliary units, such as hydraulic and pneumatic units, and giving a description of electric motors or other electrically operated units. (Figure 3.8-1)

In presenting the machine to the management, it is always good to have executives on hand from all interested departments. Questions can then be answered and settled right to the satisfaction of everyone concerned.

It is always good practice to distribute in advance copies of your concise report and specification sheet to all interested persons, especially the men who will be at the management meeting. This will give all a chance to get acquainted with the important features of the machine, and a more intelligent, fruitful discussion can follow.

3.9 Machine Model

Since some parts or features of the machine may not be clearly understandable to management, on seeing drawings and plans alone, it is wise and expedient to provide a working model. The simplest, least expensive way to get your idea across is to make a simple wooden model of the mechanism. To illustrate this, we will use an actual case, a simple patent on a geneva drive mechanism issued to the author. A geneva drive mechanism is a simple method of advancing a portion of the drive in relation to the rest of the drive and to keep this relationship for a predetermined length of time. If accuracy of advancement

3.9 Machine Model

SPECIFICATIONS
Straight Bevel Gear Machine

Capacity

Longest Cone Distance _____
Maximum Ratio _____
Minimum Ratio _____
Largest Pitch Angle _____
Smallest Pitch Angle _____
Largest Pitch Diameter _____
Coarsest Pitch _____
Longest Face _____
Index Range (number of teeth) _____
Work Piece Spindle
Diameter of Tapered Hole at Large End _____
Taper per Foot _____
Depth of Taper _____
Diameter of Hole Through Spindle _____
Speeds and Feeds
Cutter Speed (feet per minute) _____
Feed (seconds per tooth) _____

Drive

Main Drive Motor _____
Hydraulic Pump Motor _____
Cutting Oil Pump Motor _____

Miscellaneous

Floor Space _____
Net Weight _____
Gross Weight _____

Figure 3.8–1. Specification sheet.

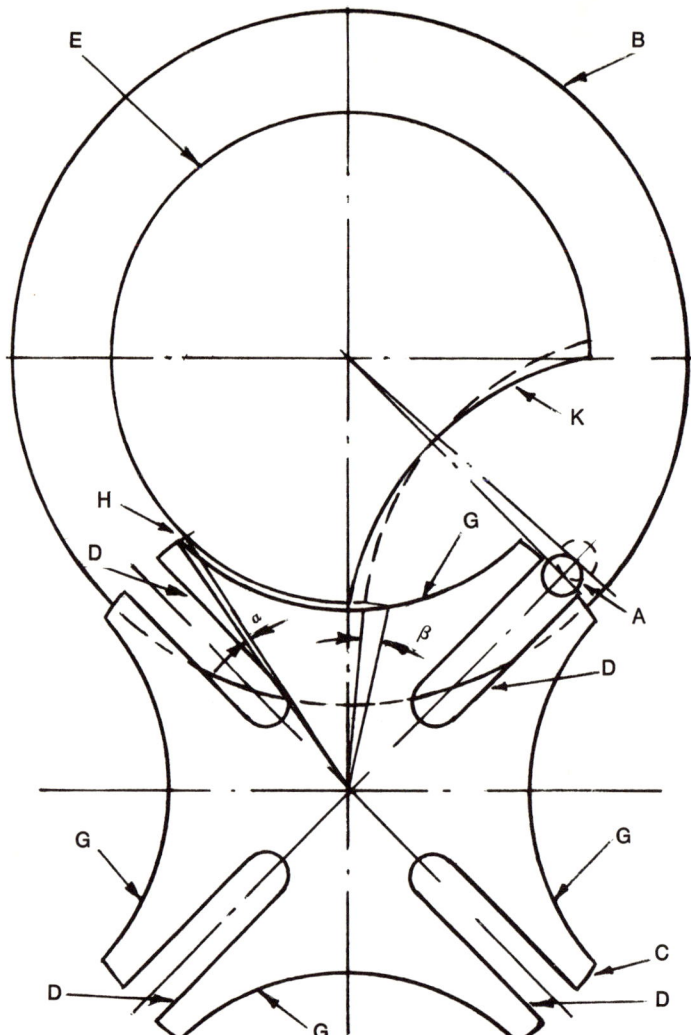

Figure 3.9–1. Conventional geneva drive. Pin A in driver plate B indexes plate C one fourth turn, driving in slot D. When plate C is not being driven, it is kept from turning by circular hub E on near side of plate B and circular portion G of plate C. When circular portion G or plate C is fully covered by hub E of plate B, the tendency of motion of plate C is 90 degrees to hub E on plate B, and it can only move the amount of the clearance H.

When, however, pin A of plate B is fully disengaged from slot D of plate C, plate C can move several times the clearance H as illustrated by phantom lines, showing position of circular index clearance K in hub E, and pin A in plate B. The possible motion of plate C could then be α in one direction and β in the opposite direction. The construction of this mechanism has been shown with sharp corners to make it easier to illustrate. With the proper radii, which is a necessary requirement for practical purposes, possible angular motion of plate C when pin A is disengaged, would be considerably more.

3.9 Machine Model

must be maintained until the next index, other means must be provided in the drive in addition to the geneva mechanism. If, however, it is not important to maintain accuracy, but is desirable to keep it reasonably close, as merely the motion from one position to the next in automatic machinery, the geneva invention by the author will meet this condition. The operation of a conventional geneva drive can easily be seen in Figure 3.9–1. The operation of the author's geneva drive, as illustrated in Figure 3.9–2, is not quite as easy to follow. The simplicity of the design, can, however, easily be demonstrated with a wooden model.

Here, a designer can clearly see in Figure 3.9–1 the great inaccuracy possible with a conventional geneva drive, and you will also readily spot the greater accuracy possible with the new drive method. This simplicity may not at all be obvious to untrained people, but they would have no difficulty seeing this when a model is available.

A good pattern maker can build an inexpensive model from simple layouts. The layouts should be full size, preferably made on inexpensive brown paper. There should be enough detail to reveal outstanding features of the machine. If the layouts are reasonably close, no dimensions are necessary for the model. The pattern maker can easily measure the layouts as he goes along. Only in cases where there may be some important motion or holding devices that you wish to show, may it be necessary to give dimensions.

A variety of methods and materials can be used in building a model, depending on the complexity of the machine, how much of the operating motion you wish to show, how long you intend to keep the model, and other factors. The frame and major assemblies of the machine can be made of commercial lumber in standard sizes, dowels and rectangular stock, and can be covered with inexpensive brown paper. If you have a sample of the part to be finished on the machine, it can be fastened to the work piece spindle. You may also have a cutter to be fastened to the cutter spindle. Perhaps some standard parts required for the model can be borrowed from the stock room, such as electric motors, push buttons, lights and gages. The model can then be painted the desired color. (Figure 3.9–3)

3.10 Reappraisal of Preliminary Design

In building the model, you may have found that to change certain features of the machine would improve its appearance or make some

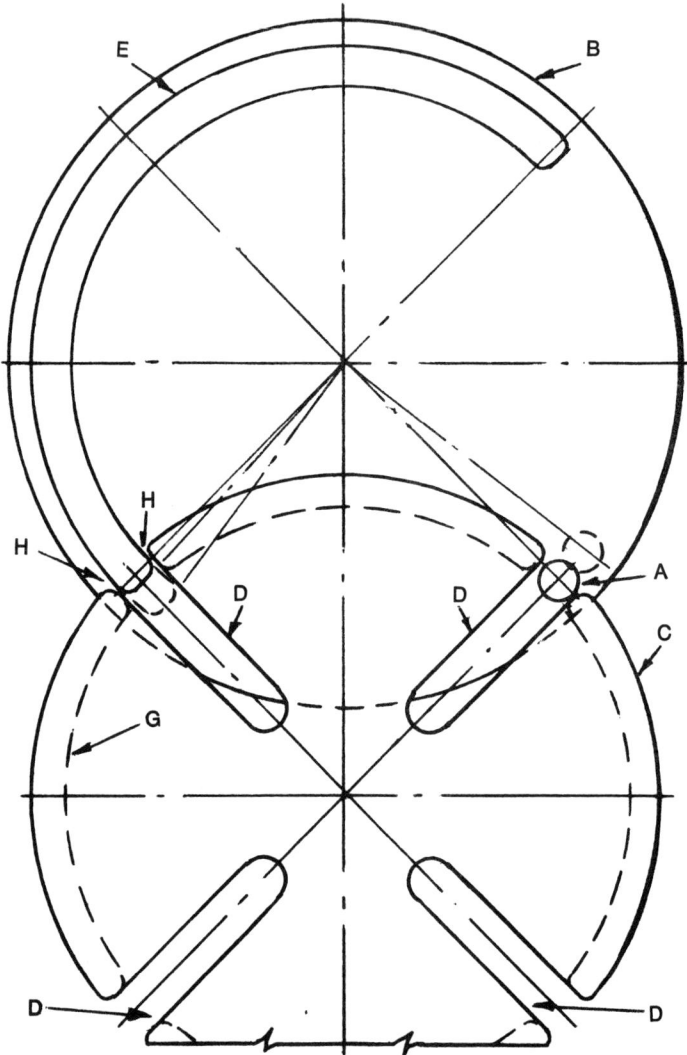

Figure 3.9–2. Author's geneva drive. Pin A in driver plate B indexes plate C one fourth turn, driving in slot D. When plate C is not being driven, it is kept from turning by ridge E on near side of plate B engaging slot D of ridge G on far side of plate C. Ridge E and G are equal in height and there is just sufficient running clearance between ridge E and plate C and ridge G and plate B. With this design there is a definite interlock of the mechanism at all times. As may be seen, pin A has already entered slot D, ready to drive plate C. Ridge E is still in slot D of plate C keeping this plate from moving. Clearance H between ridge E and sides of slot D in ridge G need only be a small running clearance. As may be seen, plate C can only move the amount of this clearance when ridge E is engaged, because the tendency of motion of plate C is always 90 degrees to ridge E.

3.10 Reappraisal of Preliminary Design

important parts of the machine more accessible without impairing operation or construction. Such changes should be made.

Photographs should be made of the full size model, and of the wooden working model, in this case the geneva drive mechanism (Figure 3.9–2), making sure that no important features have been left out.

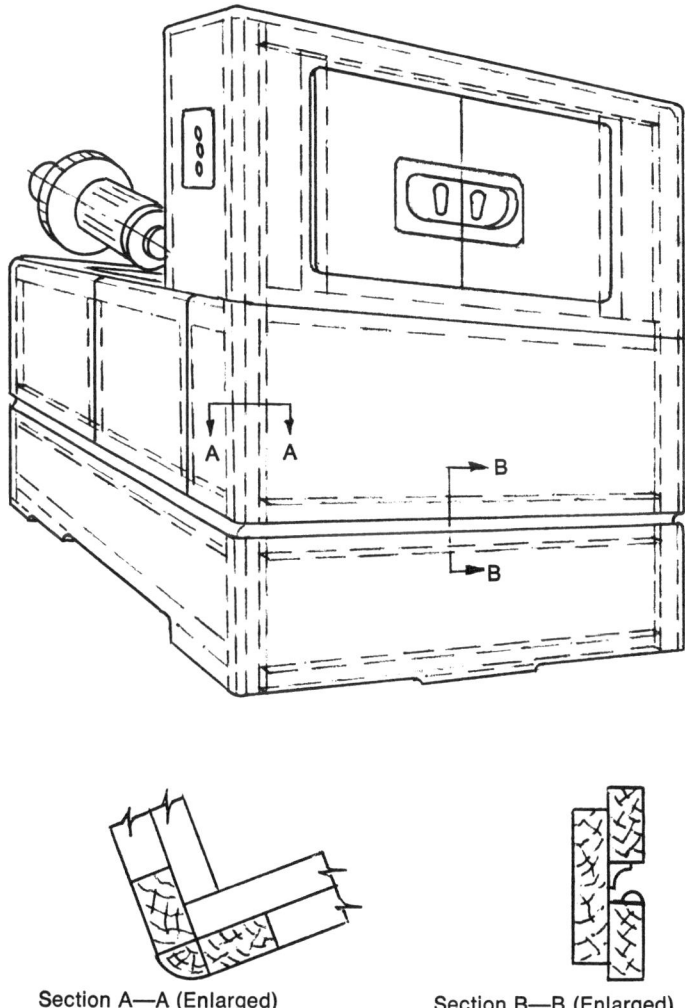

Section A—A (Enlarged)　　　Section B—B (Enlarged)

Figure 3.9–3. Showing construction of wooden model. As shown, commercial lumber may be used where very little fitting is necessary. This structure is then covered with heavy inexpensive paper, and painted the desired color. Doors may be made of thin commercial hardboard.

All interested parties can now be invited to see the models. The designer of the machine should be present at such meetings so that questions can be accurately answered. The written report covering important facts about the machine should now be brought up-to-date, and engineering superiors and management should be given copies with photographs. They are then in a position to review the situation thoroughly, and to make a decision concerning further action.

3.11 Trials and Tests

After seeing the wooden working model of the drive mechanism, engineers and management may have some questions as to life expectancy and performance. The answers can easily be found through testing. The tests can be arranged to simulate several years of service in a short time. The parts of the drive mechanism to be tested should be made and mounted with the same accuracy that you expect in the finished machine, and the type and hardness of material, and the method of lubrication should be exactly as in the finished machine. (Figure 3.11-1)

Both money and time could be saved if any other necessary tests could be made at this time.

3.11 Trials and Tests

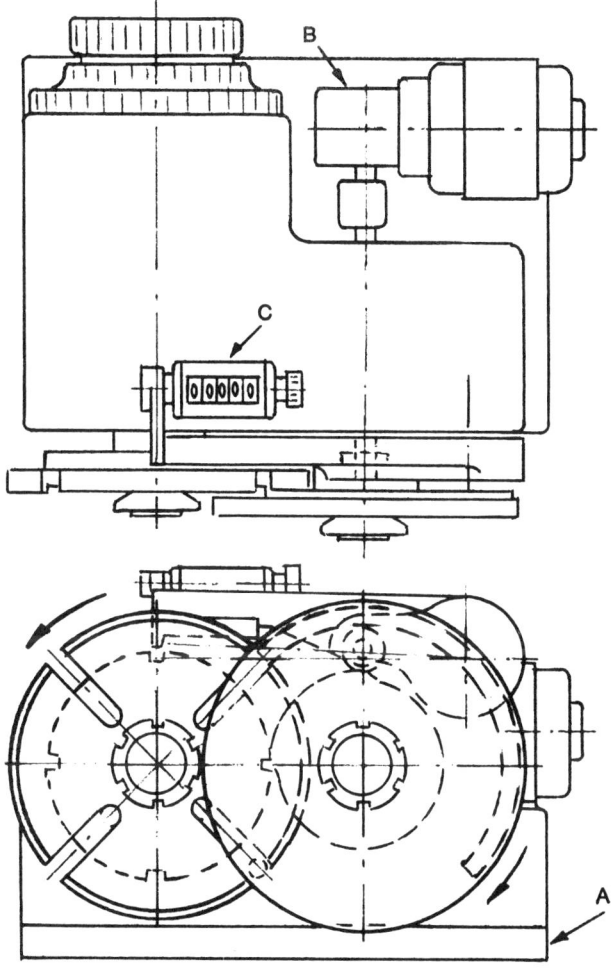

Figure 3.11–1. Life expectancy and performance test.

The unit to be tested may be mounted on a boiler plate A. A gear motor B is shown driving the unit continuously. Since in actual service this unit is running intermittently, a long time service may be simulated in a short time by running continuously. A counter C keeps record of the number of indexes, being operated by the index lockup lever moving out of engagement.

CHAPTER 4

Prototype Machine

4.1 Introduction

The machine designer may be called upon to design a variety of machine tools, the functions of some of which are listed below:

1. Cutting
   a. Boring, Turning and Facing (Lathes)
   b. Boring Mills
   c. Drilling and Boring
   d. Shapers
   e. Planers
   f. Milling
   g. Gear Generators
   h. Gear Hobbing
   i. Gear Cutting
2. Broaching
3. Grinding
4. Honing
5. Lapping
6. Heat Treating
7. Forming
8. Coining
9. Piercing
10. Calendering
11. Measuring
12. Testing
13. Textile
14. Assembling
15. Processing
16. Business
17. Domestic

In discussing the prototype machine, we will consider elements and functions of several of the machines mentioned, and give examples of successful application.

Let us assume now that you have a favorable arrangement for designing the new machine, including several designers or layout men working with you under your close supervision to lay out the various assemblies of the machine. This is an ideal arrangement which leaves you free to look into all problems, and to work out the best design.

4.1 Introduction

You have already divided the machine into sections that can easily be assembled as units and then finally joined to make the finished machine.

The men working with you on the project should understand from the beginning that you expect them to follow your preliminary design. However, they should understand that you expect them to use their own initiative to work out problems as they come up. You, on the other hand, should consider any suggestions they may have to improve the design.

In your preliminary design, you worked out an overall picture of the machine, and decided on major drive arrangements. You may also have thought of innovations that you knew would need close attention when finally designed, so that they would work harmoniously with the rest of the machine. Therefore, when working out the final design, make several full size layouts showing different arrangements of the important functions and parts of the machine. These layouts should only be simple line drawings to set you on the right course. You probably covered this in freehand sketches when you worked on the preliminary design, so it is now just a matter of drawing it up accurately to full size. Even after you have chosen the design that looks most promising, you may run up against problems that seem impossible to solve. If you do not see a solution within a reasonable time, work on some other part of the machine. Usually a good solution will come when you start to look into this design again.

Several problem areas in general are always present in the prototype design stage and should be given due consideration:

1. Lubrication should be kept in mind from the very beginning.
2. The machine should be easily accessible for the machine operator and maintenance men.
   a. The controls should be within easy reach and arranged so that they cannot be accidentally released.
   b. The height of the work piece from the floor, should be convenient for the average person in standing or seated position. (See figure 4.1-1)
3. Entrance and exit of the work pieces should be convenient.
4. Disposal of waste material should be arranged for.
5. Location of electrical components and compartments should meet standards required by industry.

Prototype Machine

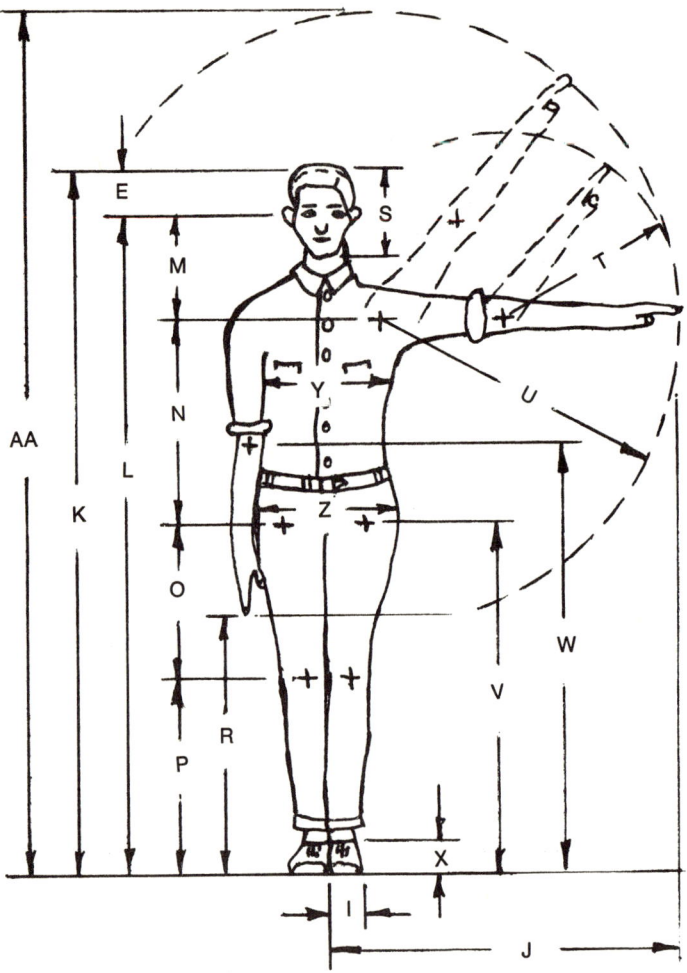

Figure 4.1-1. Outline of average American woman or man in seated or standing postion. All measurements are in inches. These dimensions will give you the most convenient locations for working areas and operating controls.

6. Since floor space is at a premium in all plants, keep the machine as small as possible without impairing its performance.
7. If the machine is to be used in a foreign country, the electrical components must conform to the requirements of that country. If the voltage and cycles are different, the rpm of the motors will be different and mechanical changes will have to be made. Gages, instruments and measuring equipment must be cali-

4.1 Introduction

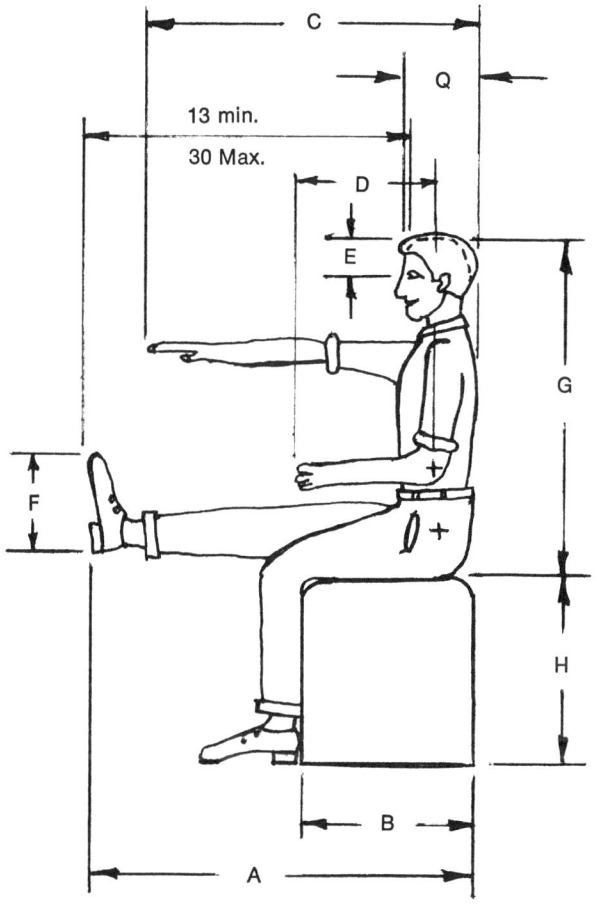

SYMBOL	WOMAN	MAN	SYMBOL	WOMAN	MAN
A	40.0	43.0	O	17.0	17.2
B	18.3	19.7	P	20.0	20.7
C	32.6	37.2	Q	7.2	7.6
D	13.0	14.5	R	25.4	23.5
E	4.0	4.5	S	8.5	9.0
F	9.8	11.0	T	13.6	17.5
G	33.9	34.2	U	28.6	29.5
H	17.0	17.0	V	37.0	37.7
I	3.8	4.0	W	39.7	43.9
J	35.6	36.0	X	3.5	3.3
K	68.0	68.9	Y	14.0	14.5
L	64.0	64.4	Z	14.9	14.3
M	10.0	8.5	AA	82.3	85.4
N	17.0	18.0			

Prototype Machine

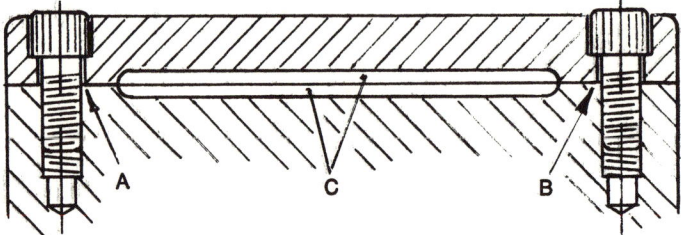

Figure 4.2–1. Correct preparation of two machine members to be fastened together permanently. The bearing surfaces under the fastening means are narrow, but large enough to support the maximum load. The machine members can either be relieved in the casting, or by rough machining. With this method, much less time is required for finish machining and warpage is not so apt to occur when finish machining, since warpage is usually due to stresses induced in the center portion when machining there.

brated for units of measure used in the country where the machine will be used.

4.2 Bearings in Machine Tool Design

The bearing or bearings may support stationary, oscillating, rotating or reciprocating members. In all cases the location, type and size of bearings are important to the proper functioning of the parts.

Stationary Bearings

We may, for instance, have a stationary bearing, a bearing to which two machine members are permanently fastened together. (See figure 4.2–1) Here the two members are fastened together at A and B, which are the bearing surfaces. These surfaces are far enough apart for good control, and of sufficient size to support the required loads.

If proper considerations had not been given for bearing surfaces, we could have had a condition as shown in Figure 4.2–2 or 4.2–3. Two surfaces finished commercially, may both have good qualities as far as surface finish is concerned. There may, however, be slight inaccuracies due to warpage when machining, or to some other cause, so that at some point there may be a seemingly insignificant distance (C in Figure 4.2–2 or D in Figure 4.2–3) which would cause the two surfaces, when under load, to vibrate in relation to each other.

Extreme care must be used to locate the fastening means so that the two surfaces make proper contact when the fastening force is applied.

4.2 Bearings

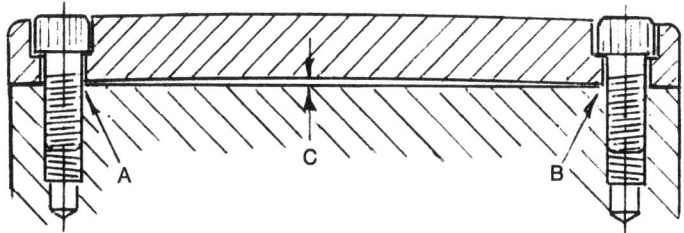

Figure 4.2-2. Wrong preparation of two machine members to be fastened together permanently. The bearing surfaces under the fastening means should be narrow, but large enough to support the maximum load. If the finished surfaces have not been relieved, either as cast or by rough machining, unnecessary time and effort must be spent in machining the surfaces to keep them from warping. This drawing shows a warped member, and since the leverage from the edge of the member to the center of the screw is short, the screw cannot be tightened down to get surface contact and the member would vibrate under load, causing the screw to become loose.

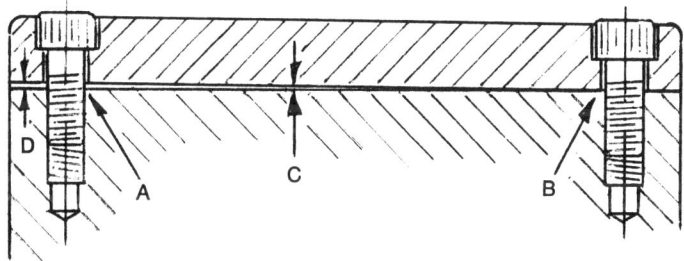

Figure 4.2-3. Wrong preparation of two machine members to be fastened together permanently. The bearing surfaces under the fastening means should be narrow, but large enough to support the maximum load. If the finished surfaces have not been relieved, either as cast or rough machining, unnecessary time and effort must be spent in machining the surfaces to keep them from warping. This drawing shows a warped member. The screw has been tightened down at B for surface contact. The member may, however, be stiff enough to prevent tightening the screw at A for surface contact. Under load, the member would vibrate and cause the screw at A to become loose. It is also conceivable that contact would be made at C and neither of the screws could be tightened. In this case, both of the screws would work loose due to vibration under load.

In Figure 4.2-2 there may be contact between the two surfaces at A and B, but only line contact under load. Since a long distance exists where there is no contact between the two surfaces at C, the slightest vibration of the structure may cause a gradual breakdown of material at the line of contact, eventually causing looseness.

In Figure 4.2-3 a fastening force has been applied which is in-

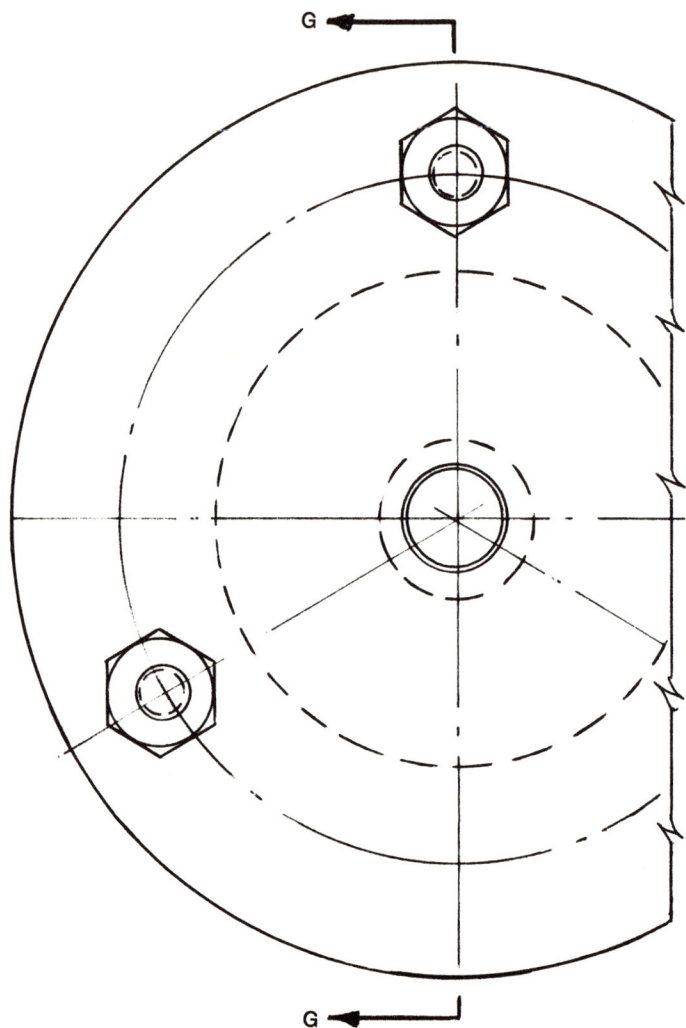

tended to be of sufficient magnitude to bring the two surfaces in contact at A and B. The parts are, however, stiff enough to prevent contact between the two surfaces at A. Under variable load, the fastening medium would gradually stretch and eventually become ineffective.

Bearings Under Motion

For rotating, oscillating or reciprocating members, more than two bearings per member are not justifiable unless special precautions are

4.2 Bearings

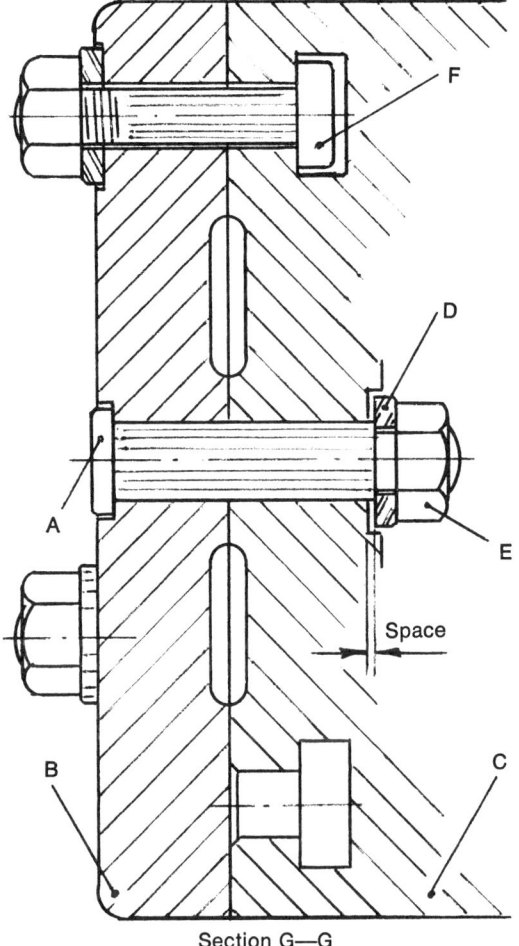

Figure 4.2–4. Simple method for centering an adjustable, circular slide. There are, of course, many variations to this example. Basically, however, this is correct, because nut E can be fastened permanently against washer D to shoulder of centering stud A, thus permitting adjustment without touching the centering stud. Notice that nut E is concealed so it cannot be mistakenly loosened.

taken to prevent the extra bearings from "fighting" the others. Of course, there must be good, valid reasons for the added expense. In many cases, anti-friction bearings are close together, and are arranged to provide thrust in both directions. Therefore, they can be considered as one bearing, and a third bearing becomes necessary for good control.

Many factors determine the selection of type of bearing: speed,

Prototype Machine

accuracy, rigidity, forces involved, heat transfer, lubrication available, contamination of processed parts, cost.

The proper bearing selection can be made only after analyzing all existing conditions and requirements. It is not unusual to change the choice of bearing after the design has made considerable progress. There may even be a definite preference of one type to another, which would become evident only after the design has been worked on for some time.

Some of the basic bearings under motion are plain, ball and roller.

Plain Bearings

Proper application of each type of plain bearing should be thoroughly understood to obtain the best results. The simplest plain bearings are two flat surfaces under motion with respect to each other. This motion may merely be an adjustment of one part in relation to the other, or a slow continuous or intermittent rotating, oscillating or reciprocating motion of one part in relation to the other. It could, of course, also be two or more parts under motion in relation to each other and a stationary member. Each type of motion requires analysis on the basis of which the proper choice of bearings can be made.

Figure 4.2–4 shows a circular slide. Whether this slide is mounted in a horizontal, angular or vertical position, how the forces acting on the slide are applied, and what the speed of motion is, or whether the slide is just an adjustable member determines how the slide should be guided and centered.

Assume that it is an adjustable member. The simplest application would then be, if weight is no problem and maximum rigidity is desired, cast iron material centered with a soft steel stud. The stud A would then be a metal to metal fit in part B, and a push or running fit in part C. (These expressions for fits are no longer in use on detail drawings where definite dimensions and tolerances are now applied, but are merely expressions signifying the class and type of fits.) Notice that washer D is drawn down tight to shoulder stud A with nut E. The dimensions and tolerances are such on parts A, B and C that a small clearance will result between washer D and part C. The slide can therefore be mounted in any position and will remain assembled when bolts F are loosened and adjustment is made. This is an inexpensive design because the center hole in part B can be bored in the same setup in which holes for bolts F are drilled. The slide surface for part B

4.2 Bearings

is also easily finished with no projections. These are important considerations in the process of designing.

Let us now assume that this slide should be designed for power drive under slow motion. Then a simple arrangement as shown in figure 4.2–5 could be designed. Here pin A is the centering means, and since this is a slow motion power drive, the pin could be made of soft steel. The pin is a press fit in part B and a running fit in bushing D. The bushing may be made of cast iron or a copper alloy, depending on the importance of the drive. A pin is chosen for centering the slide, rather than depending on the large diameter of the slide for centering. Some of the reasons for this selection are:

1. Preparation. It is much easier and much more economical to finish a small diameter pin and a small diameter hole to close tolerances for a close central fit than to finish two large internal and external diameters to the same accuracy.
2. Lubrication. It is also easier to lubricate a small diameter pin. With this arrangement, axial thrust surfaces of member C can be lubricated by leading the lubricant into space F at outer diameter of member C.
3. Force and Friction. In analyzing the magnitude of friction resulting from the force applied to turn member B (if this force must be applied between the center and space F or even a small distance outside space F) you will find that the only solution would be to use a pin for centering, unless the design of the whole structure would warrant the use of anti-friction bearings on the large diameter.

The plain bearings for motion may be subdivided as follows:

    a. Cast Iron
    b. Nonferrous Metal
    c. Powdered Metal
    d. Metallic Graphite Suspended
    e. Carbon
    f. Plastic Material
    g. Precious Stone

CAST IRON PLAIN BEARINGS. Cast iron is one of the most common materials used in machine tool design. It is, therefore, not unusual to see a steel shaft or rod supported in holes bored directly in a cast iron

Prototype Machine

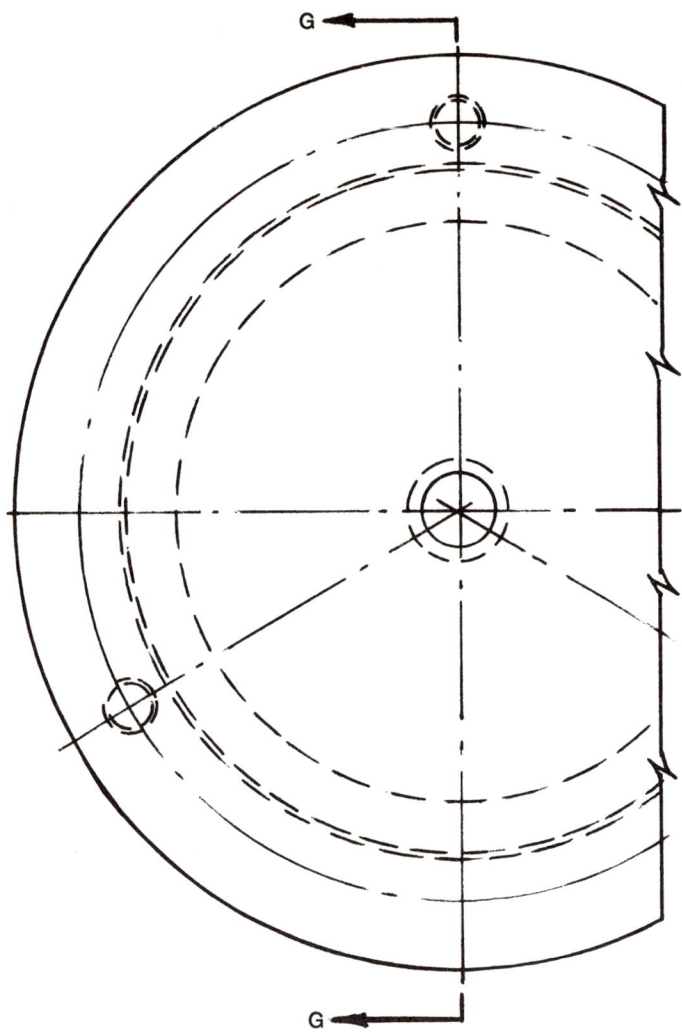

machine member. This is acceptable for adjustable members if lubrication is available. It could also be acceptable for a slow, lightly loaded power drive if an ample supply of lubricant is available, and if it can be guaranteed that the required accuracy can be maintained throughout the expected life of the machine. The life expectancy of cast iron bearings for a smooth steel shaft is excellent for slow speeds and light loads when the bearings are well lubricated. If, however, there is the slightest doubt as to life expectancy, and a plain bearing otherwise is

4.2 Bearings

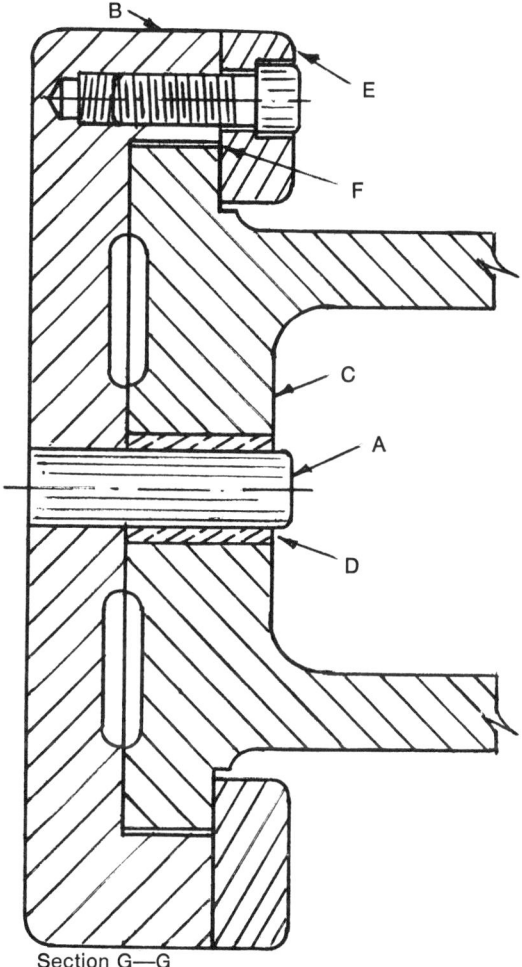

Section G—G

Figure 4.2–5. Simple method for centering a circular slide for power drive under slow motion. There are, of course, many variations to this basic solution. One thing should, however, be kept in mind. If the circular surface at F were used as a bearing diameter, and if the driving force to rotate slide B were applied anywhere between the center and surface at F, or even a small distance outside F, with a small radial load on slide B, the frictional forces would be too great. Then the only solution instead of a plain bearing at F would be an anti-friction bearing or, as is shown, a central small diameter pin and plain bearing.

acceptable, bushings, which are easily replaceable, should be used. The best policy is, however, to select bearings that will have a life expectancy at least as good as that of the rest of the machine.

Cast iron is also a good plain bearing material selection for two flat surfaces in sliding contact for adjustment and power drive at slow and moderate speeds. In either case, oil grooves should be provided in one member to be sure that every part of the two surfaces has an ample supply of oil. If the two surfaces are sliding or turning at higher speeds, proper lubrication is very important.

NONFERROUS METAL PLAIN BEARINGS. This class of bearings consists usually of an alloy of two or more nonferrous metals. Some of the most common alloys are bronze or copper, Babbitt or White Metal and aluminum.

These alloys are adequately described in other handbooks and manufacturers' catalogs. We will mention only that the most commonly used compositions for machine tool application are SAE #660 bronze for slow or medium speed applications, and SAE #65 phosphor bronze for higher speeds and heavier load applications.

These bearings are often used for power drives of slow or moderate speeds where ample space is available for good control. They must be provided with the proper oil grooves, and have an adequate supply of lubricant. Life expectancy is very good when they are in contact with a smooth steel shaft or rod in the soft or lower hardness range. For accurate drives, the design must be so arranged that they can be line reamed at assembly.

POWDERED METAL PLAIN BEARINGS. These bearings are sold under several trade names and are used to support rods and shafts for adjustments, or, less important, for slow, light-load, power drives, especially in inaccessible locations where supply of lubricant from a central source is not available. If most of the lubricant has disappeared, as happens sometimes during storage, the bearings must be soaked in a lubricant before installation. These bearings should not be pressed down to a shoulder, which might damage the bearing. Make sure that when the bearing is assembled flush there is plenty of room between the end of the bearing and the shoulder. A special sizing tool is usually required for proper installation. (See figure 4.2–6)

METALLIC GRAPHITE SUSPENDED PLAIN BEARINGS. These bearings are used where lubrication is a problem, where good quality is expected and where a higher hardness is necessary than that possible with the common bronze bushing.

CARBON PLAIN BEARINGS. These bearings are used where high temperature exists and where lubrication is a problem.

4.2 Bearings

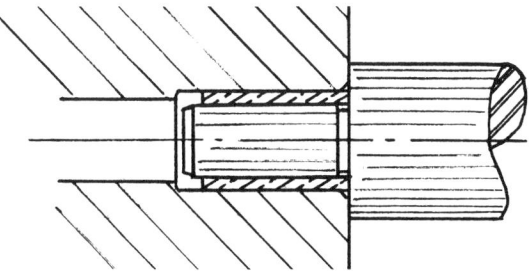

Figure 4.2–6. Installation of powdered metal plain bearings. The design must be worked out in such a way that the bearing will not be pressed in against a shoulder. This would damage the bearing. This type of bearing is installed with a sizing tool to obtain the correct bearing clearance.

PLASTIC MATERIAL PLAIN BEARINGS. These bearings are used where lubrication is a problem, and where speeds and loads are not too high. If speeds and loads are too high, certain plastic materials will cold flow, and in some cases will disintegrate. Many plastic materials are recommended for use with no lubrication. They are well suited for food processing or textile machinery where contamination can not be tolerated, and when the loads and speeds are moderate. If the medium being processed in the machine is in a liquid form, this may sometimes help to lubricate the bearing surfaces.

PRECIOUS STONE PLAIN BEARINGS. These bearings are used mainly for instruments and in parts of a machine where high accuracy is required. Sapphire is a very acceptable material.

Generally, it should be noted, that all plain bearings, even if manufacturer specifies that they will work dry, give the best performance when lubricated. The lubricant may not always be an oily substance, but may be other liquids or gases, as found in food processing or medicine processing machines.

Ball Bearings

There are two main categories of ball bearings, ground and unground.

Ground Ball Bearings

The ground ball bearings may be subdivided into the following groups:

a. Radial Load Conrad

b. Radial Load Higher Capacity, with Loading Grooves
c. Angular Contact
d. Double Row
e. Instrument or Miniature
f. Slim Line

RADIAL LOAD CONRAD BALL BEARINGS. This type of bearing is widely used in machine tool design, and will satisfy a great number of designs to support radial loads and also thrust loads in two directions. The balls are loaded between the outer and inner rings by displacing the inner ring eccentrically. The number of balls per bearing are thus limited to the available opening space. The bearing rings are uninterrupted. This is an important consideration in some cases where extreme accuracy and high speeds are required, calling for a ring of uniform section. (Figure 4.2–7)

RADIAL LOAD HIGHER CAPACITY BALL BEARINGS WITH LOADING GROOVES. This bearing is similar to the conrad bearing. However, it

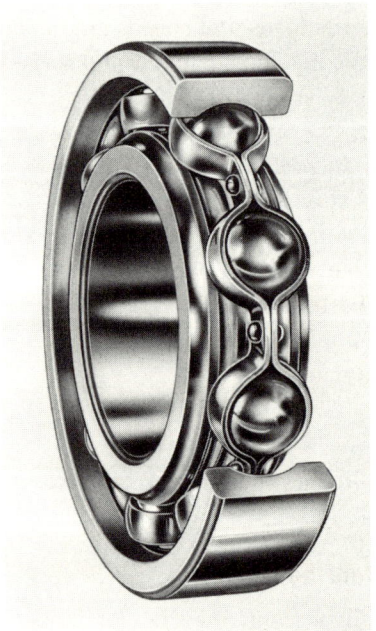

Figure 4.2–7. Radial load conrad ball bearing. (Courtesy of New Departure-Hyatt Bearings Division of General Motors Corporation)

4.2 Bearings

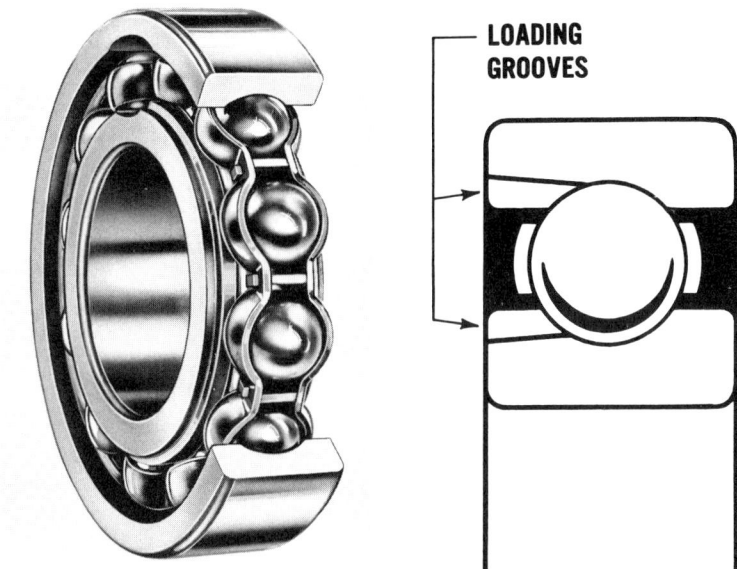

Figure 4.2–8. Radial load maximum capacity bearing with loading grooves. (Courtesy of New Departure-Hyatt Bearings Division of General Motors Corporation)

does have a filling slot or loading groove in the outer and inner rings, permitting the maximum number of balls to be assembled in the bearing and providing a greater capacity for the same size bearing. This is a high accuracy bearing for high speed applications. When extreme accuracy and exceptionally high speeds are encountered, it would be well to consider that these bearings are of a nonuniform section circumferentially due to the loading slot, and may cause some problems. (Figure 4.2–8)

ANGULAR CONTACT BALL BEARINGS. These bearings are widely used in machine tools, especially for high precision spindles and shafts. The bearings can be furnished by the manufacturer in duplex pairs, thus providing for the highest in load capacity, accuracy and rigidity. (Figure 4.2–9) The duplex pair would, as a rule, be located close to the applied load for greatest accuracy, and a third bearing, which could be located at the other end of the shaft or spindle for good control, could be a radial ball bearing or a straight roller bearing.

Prototype Machine

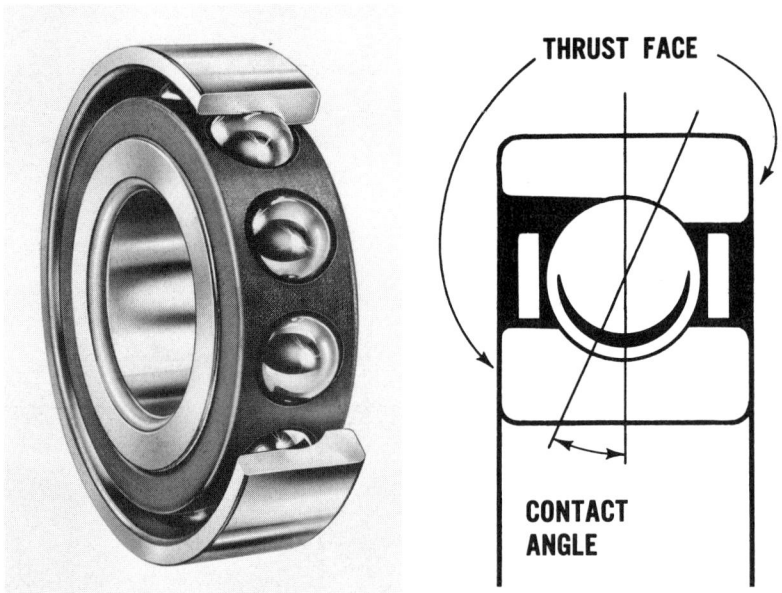

Figure 4.2–9. Angular contact ball bearings. (Courtesy of New Departure-Hyatt Bearings Division of General Motors Corporation)

Preloaded Duplex Ball Bearings

To reduce the deflection of a bearing when under load, the principle of preloading has come into general use. An axial preload is a controlled thrust load which is applied to a pair of bearings by certain design and mounting techniques. Bearing preloading has particular significance in machine tools and other fields where spindle rigidity is of paramount importance.

Preload is usually accomplished in a bearing application by one of the following methods:

1. Use of a pair of ball bearings with controlled offset or stickout between inner and outer ring faces. When mounted DF or DB so that both inner and outer ring faces of a pair of bearings contact, a predetermined axial preload results. This effect can also be accomplished by placing equal length inner and outer ring spacers between bearings and clamping all parts to achieve abutment.

2. Use of a nut adjustment on the shaft or housing which, when

4.2 Bearings

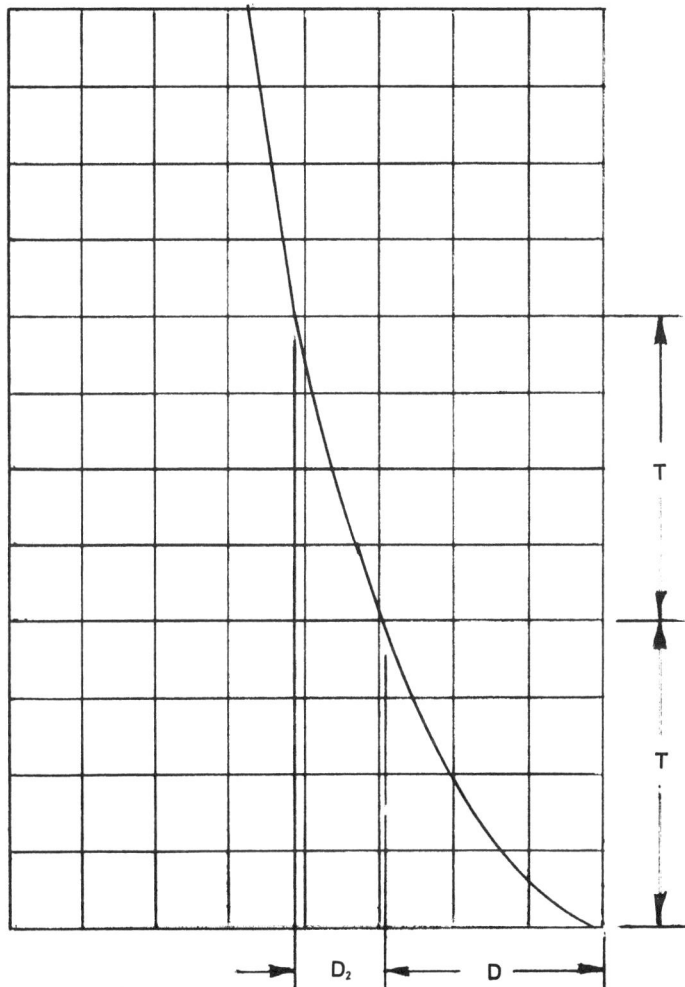

Figure 4.2–10. Axial deflection versus axial load.
T = Axial load
D = Initial axial deflection due to axial load T
D_2 = Axial deflection by an additional axial load T of same magnitude
(Courtesy of New Departure-Hyatt Division of General Motors Corporation)

tightened, displaces one bearing ring axially and introduces the desired axial preload.

3. Use of a thrust washer or springs to apply the necessary axial thrust.

The first method is relatively simple and automatic with success

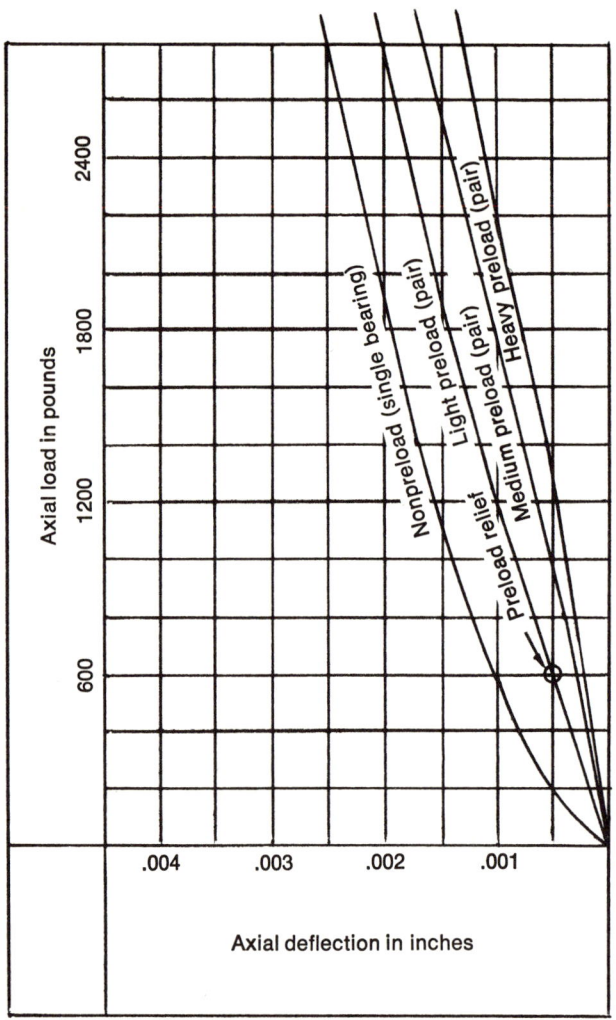

Figure 4.2-11. Relief of preload. (Courtesy of New Departure-Hyatt Bearings Division of General Motors Corporation)

depending primarily upon the accurate manufacture of the bearings. The second depends for success upon skill and experience in bearing installation. The third method requires judgment in designing the correct preload.

AXIAL DEFLECTION VERSUS AXIAL LOAD. A typical curve of axial deflection versus thrust load for a representative single row angular contact bearing is shown in Figure 4.2-10. The initial axial deflection

4.2 Bearings

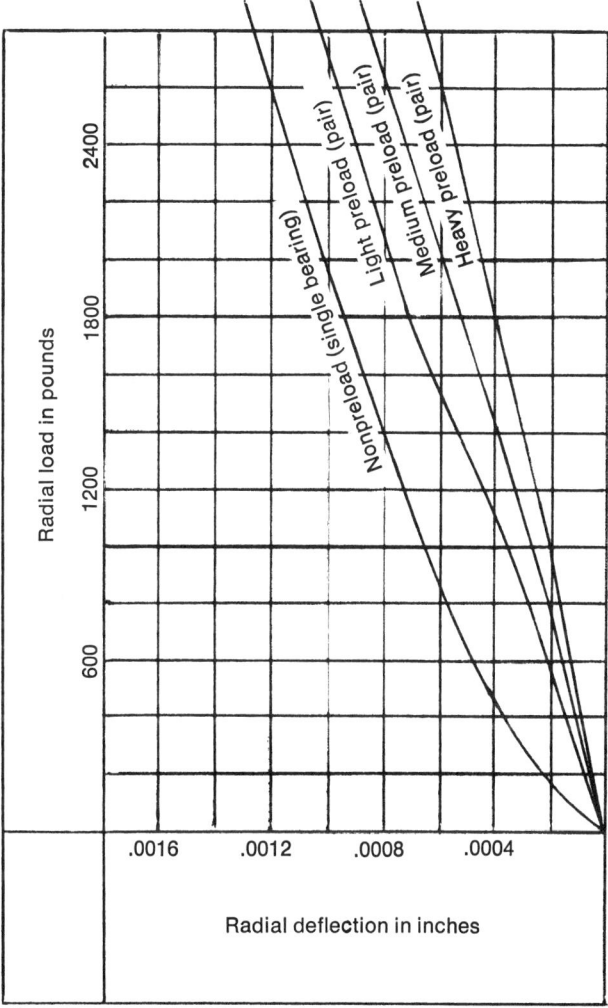

Figure 4.2–12. Radial deflection versus radial load. (Courtesy of New Departure-Hyatt Bearings Division of General Motors Corporation)

caused by a load T is much greater than the further deflection caused by an additional load of the same magnitude. It is apparent from this that if the first or lower part of the curve were to be eliminated before the work load is applied closer control of deflection could be obtained. This is accomplished in ball bearings by preloading.

The reduction in axial deflection obtained by preloading a typical duplex pair of angular contact bearings is indicated in Figure 4.2–11.

Prototype Machine

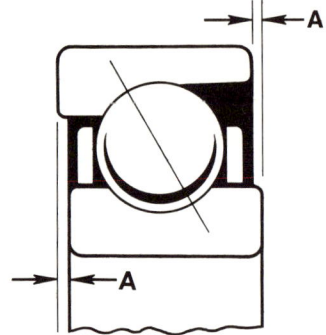

Figure 4.2–13a. Single bearing showing equal face stickout on either side.

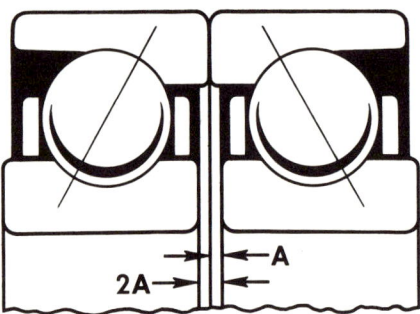

Figure 4.2–13b. DB mounting before clamping. When inner rings are abutted, a preload corresponding to an axial deflection of A will exist.

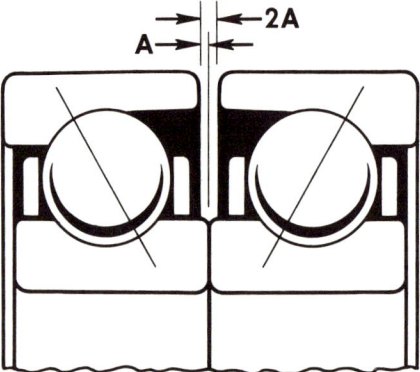

Figure 4.2–13c. DF mounting before clamping. When outer rings are abutted, a preload corresponding to axial deflection of A will exist.

4.2 Bearings

Figure 4.2-13d. DT mounting in which both bearings share thrust load. DT pairs may be preloaded against single angular contact bearings or duplex DT pairs. (Courtesy of New Departure-Hyatt Bearings Division of General Motors Corporation)

Curves are given for nonpreloaded single bearing and for duplexed pairs having light, medium and heavy preloads.

RELIEF OF PRELOAD. When an external thrust load is applied, an additional deflection occurs in the preloaded thrust resisting bearing and results in a partial relief of the preload. Thus the bearing carries the work load plus the remaining preload. As the work load is increased, a point at which the preload is entirely relieved will eventually be reached.

To illustrate this, refer to the curves in Figure 4.2-11. The light preload for the bearing represented is taken as 200 pounds. From the single bearing curve we see that the initial deflection for one bearing is .0005 inch under 200 pounds thrust load.

For a pair of light preloaded bearings, it would require 600 pounds thrust load to obtain this deflection, a condition under which preload would be entirely relieved.

RADIAL DEFLECTION VERSUS RADIAL LOAD. When angular contact bearings are preloaded, the deflection due to radial loads is reduced as indicated by the typical curves in Figure 4.2-12. Note the characteristic change in deflection rate when preload is relieved.

Duplexing

Duplexed pairs of bearings are furnished with a specially ground face relationship which will result in the predetermined preload when the bearings are mounted with opposed contact angles.

As shown in Figures 4.2-13a to 4.2-13d, the individual bearings

Prototype Machine

are made so that the faces of the inner and outer rings will be flush on both sides under the predetermined internal axial force. This permits the two bearings of a duplex pair to be mounted in any of the duplex mounting arrangements, DB, DF, or DT as shown in Figures 4.2–14a to 4.2–14c.

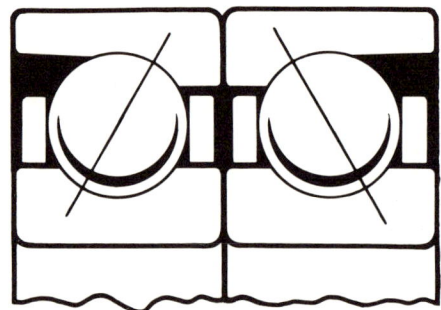

Figure 4.2–14a. Duplex DB back to back mounting.

Figure 4.2–14b. Duplex DF face to face mounting.

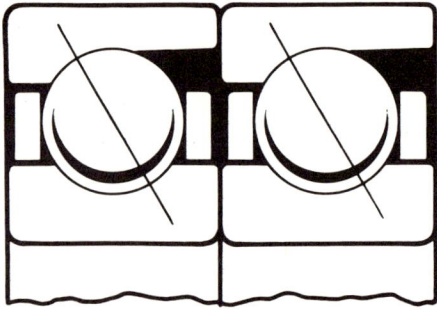

Figure 4.2–14c. Duplex DT tandem mounting.
(Courtesy of New Departure-Hyatt Bearings Division of General Motors Corporation)

4.2 Bearings

The amount of race offset represented by A in Figures 4.2–13a to 4.2–13d results from the preload grinding operation, and is equal to the axial deflection of the single bearing under a thrust load identical in magnitude to the preload.

Although the great majority of applications will be satisfied by the standard preload, it is occasionally necessary to furnish a special preload. In such cases, the manufacturer should be consulted for recommendations.

EFFECT OF MOUNTING FITS ON PRELOAD. When bearings are press fitted on a shaft, the resulting slight expansion of the inner ring increases the preload in a pair of mounted duplex bearings. Accordingly, it is important that the machine builders use the recommended shaft and housing fits when employing duplexed angular contact bearings for precision work. Otherwise, a much greater than intended preload can result.

Housings are normally bored so that the bearings are a push fit before preloading. By clamping either the inner or outer rings, depending on whether mounting is DF or DB, after the bearings are assembled in position, the resulting preload and outer ring expansion can provide a light press fit in the housing. Accordingly, additional rigidity and easier machine assembly are accomplished by proper preloading practice.

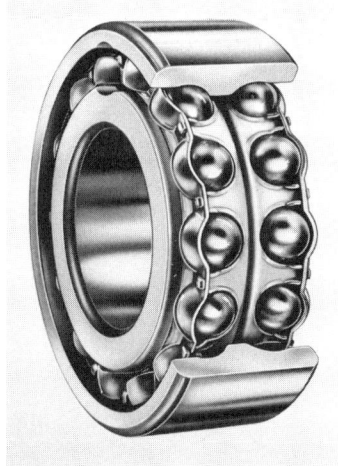

Figure 4.2–15. Double row ball bearing. Rigid type. (Courtesy of New Departure-Hyatt Bearings Division of General Motors Corporation)

Prototype Machine

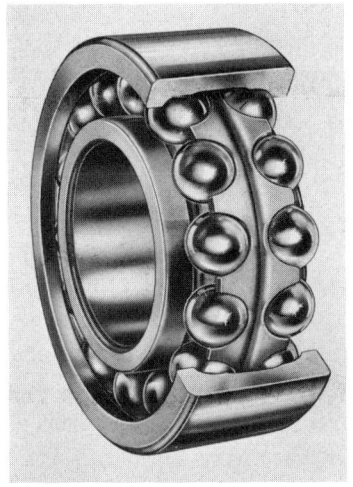 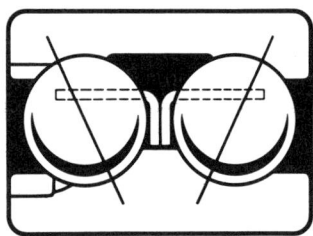

Figure 4.2–16. Double row ball bearing. Flexible type. (Courtesy of New Departure-Hyatt Bearings Division of General Motors Corporation)

DOUBLE ROW BALL BEARINGS. These bearings usually come in two basic design types, rigid and flexible.

In the rigid type the lines of contact converge outside the outer race. This bearing is useful for pulleys, idler gears or similar short shaft applications. (Figure 4.2–15)

In the flexible type the lines of contact converge inside the inner race. This bearing should not be used alone. It requires another bearing for control. It is very good for a long shaft where extreme accuracy is not required, and could then be used with a plain bearing, a needle bearing or a single radial ball bearing, depending on speed and accuracy desired. (Figure 4.2–16)

Instrument Bearings

As the name implies, these bearings are used in machine design for instruments or checking devices, or in auxiliary components. Some of the high speed types are recommended by the manufacturer for speeds in excess of 350,000 rpm. (Figure 4.2–17)

There are also many types of ball bearings for linear motion, and special applications of balls for screws and similar contrivances, which recirculate the balls in their environments. These units are useful in reducing friction. Careful consideration should be exercised in their

4.2 Bearings

Figure 4.2-17. Typical variety of instrument bearings. (Courtesy of MPB Corporation)

application for metal working machines to make sure, when heavy loads are applied, that the finished product of the machine will have the high quality expected. (Figures 4.2-18 and 4.2-19)

Unground Ball Bearings

Unground ball bearings are seldom used for machine tools, and then only for auxiliary equipment such as conveyors.

Roller Bearings

There are two main categories of roller bearings: straight or cylindrical and tapered.

Prototype Machine

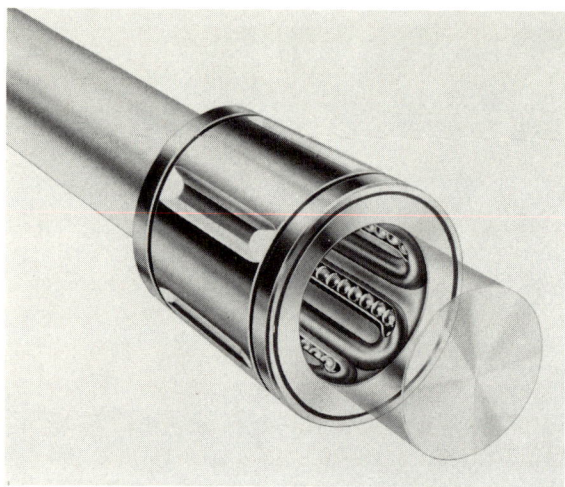

Figure 4.2–18. Ball bushing. This type of linear motion ball bearing usually has three or more circuits of balls to carry the load and maintain alignment. Two bearings, far enough apart for good control, are, however, necessary in most cases on one of the shafts, if two parallel shafts are used in the design. If two bearings are necessary to carry the load on the second shaft, they should be close together, so they will not fight the bearings which are far apart on the first shaft. (Courtesy of Thomson Industries, Inc.)

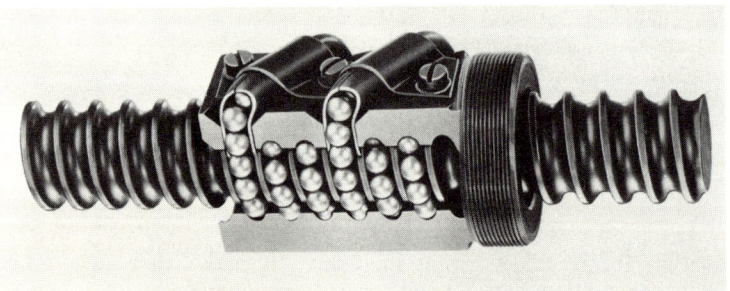

Figure 4.2–19. Ball bearing screws. This type of bearing design is very useful where it is desirable to reduce friction and eliminate stick-slip, a very perplexing occurrence when making a small amount of accurate adjustment from standstill with a plain adjusting screw. Considerable force may have to be applied before any motion takes place, and then, suddenly, too much motion.

The life expectancy of a ball bearing screw design can now be predicted fairly closely. Serious consideration should be given before definite decision is made. High, repetitive impact loads will seriously affect the life. A long, slender screw is also difficult or impossible to keep straight in the process of hardening, and if for this reason the hardness must be reduced, this would also shorten the life of the design. (Courtesy of Saginaw Steering Gear Division of General Motors Corporation)

4.2 Bearings

Straight or Cylindrical Roller Bearings

These bearings may be subdivided into the following groups:

a. Retainer Type
b. Full Complement, the majority of which are commonly called Needle Bearings
c. Roller Type and Needle Type Thrust

RETAINER TYPE ROLLER BEARINGS. These bearings are adapted to much higher speeds than full complement roller bearings having no retainer. In machine tool applications they should be used to support radial loads or in cases where no thrust loads exist and precise axial stability is of minor importance. (Figure 4.2–20) These bearings come with or without inner race rings. For high precision work, however, a bearing without an inner race ring is more practical because it saves space, eliminates the need to adapt the shaft to retain an inner race ring, and has fewer fits to be concerned with.

FULL COMPLEMENT TYPE ROLLER BEARINGS AND NEEDLE BEARINGS. There are two main types: drawn cup needle bearing (thin section), and machined cup needle bearing (heavy section).

DRAWN CUP NEEDLE BEARING. A full complement roller bearing is called a needle bearing when the bearing pitch diameter to roller diameter, and the roller length to roller diameter ratios are large. This

Figure 4.2–20. Retainer type roller bearing. (Courtesy of The Torrington Company)

Figure 4.2–21. Drawn cup needle bearing. (Courtesy of The Torrington Company)

Prototype Machine

bearing is made with an outer race ring of deep drawn, low carbon steel, with a case hardened race and a ductile core. (Figure 4.2–21) The thickness of the cup material along the circumference is accurately uniform, but because the material is thin, the roundness of the cup race ring can only be as accurate as the housing bore. Therefore, the cup should be a press fit in the housing, and the housing should be sufficiently heavy and of a uniform section to assure an accurate cup race. These bearings come with or without inner race rings. In most cases the inner race ring is omitted in machine tool design so that the shaft may be made as stiff as possible for the space available. Avoid using this type of bearing where heavy impact loads are expected.

When assembling the outer race ring in its housing, make sure that the bearing will not be seated against a shoulder when pressed flush with the housing face, for this would damage the thin section of the outer race ring.

Machined Cup Needle Bearings

These bearings come with or without inner race rings. (Figures 4.2–22 and 4.2–23) The race rings and needles are, as a rule, made of high carbon alloy steel. The race rings are ground after heat treatment, and because the section is heavy, the roundness of the rings is very close. The outer race ring is usually a light press fit in the housing bore.

Figure 4.2–22. Machined cup needle bearing with inner race ring. (Courtesy of Orange Roller Bearing Company)

Figure 4.2–23. Machined cup needle bearing without inner race ring. (Courtesy of Orange Roller Bearing Company)

4.2 Bearings

High grade needles to very close tolerances may be purchased from the manufacturer to be used in cam rollers or similar devices.

Roller Type or Needle Type Thrust Bearings

The rollers or needles in these bearings are usually separated by retainers. These bearings have a limited use in machine tool design and have greater use in auxiliary equipment and domestic machinery where a high degree of accuracy does not have to be considered. In spite of the cylindrical shape of the needles or rollers, which invites sliding as a result of the difference in supporting diameters, they operate very well under moderate loads and comparatively high speeds. (Figure 4.2–24)

Figure 4.2–24. Straight needle thrust bearing. (Courtesy of The Torrington Company)

Some straight type roller bearings are designed to carry combined radial and thrust loads. They are also applied to linear motion arrangements, but have limited use in machine tool design.

There are, however, roller applications well adapted to linear motion in which the bearings are capable of carrying heavy loads with a low coefficient of friction. Two hardened and ground shafts are required, which should be supported in the bed of the machine along its entire length for rigidity. A dual recirculating roller arrangement, V-mounted, serves as a guide and carries both the supporting load and alignment load. The single bearing arrangement on the other shaft carries load only in one direction. (Figure 4.2–25)

Tapered Roller Bearings

Tapered roller bearings are extensively used in machine tools. The

Prototype Machine

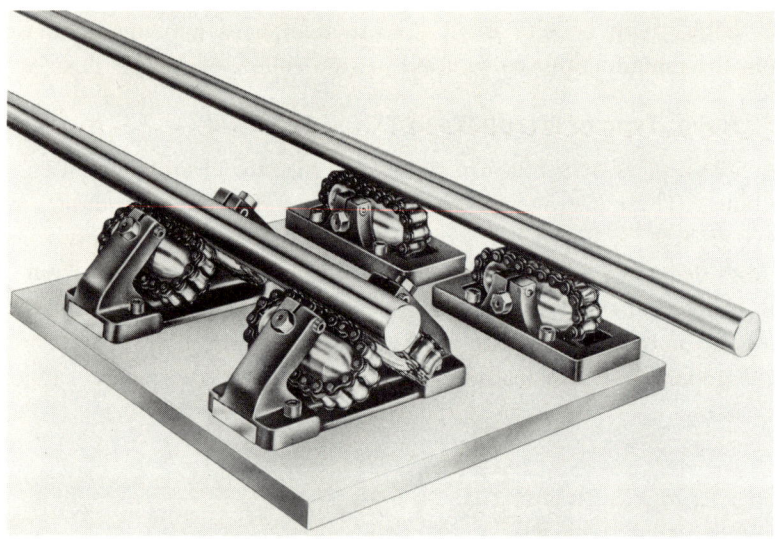

Figure 4.2–25. Roundway bearings and ways. Recirculating concave hardened and ground rollers on hardened and ground steel shafting, supported in the bed of the machine. (Courtesy of Thomson Industries, Inc.)

characteristics of true rolling surfaces are adequately illustrated in Figures 4.2–26 and 4.2–27.

Tapered roller bearings most commonly used in machine tools may be divided into three main groups:

1. Standard Type
2. Steep Angle Type
3. Thrust Type

STANDARD TYPE TAPERED ROLLER BEARINGS. These bearings are the most widely used types in machine tool design. (Figure 4.2–28) They are very well suited to carry heavy loads. Two opposed bearings must be used to carry the load and stabilize the shaft or spindle. Where speeds are high, and if one bearing were located at each end of a long shaft or spindle, sufficient heat may be generated to lengthen the shaft enough to cause trouble. If it then would be necessary to pull the two bearings together at the load end to carry most of the radial and all of the thrust load, a third bearing would have to be chosen to carry pure radial load at the other end of the shaft or spindle. It would then have to be determined whether space, cost and other factors would permit this design.

4.2 Bearings

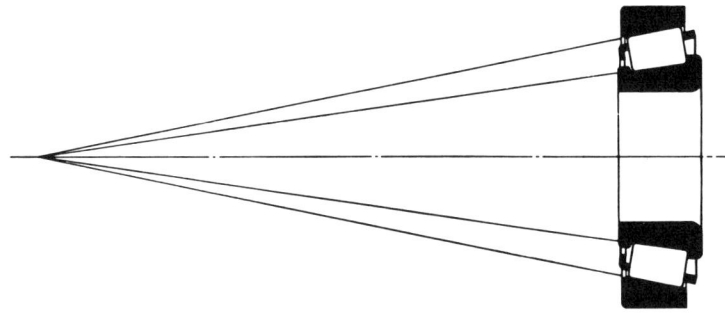

Figure 4.2–26. The basic principle of tapered roller bearing. The apexes of the tapered surfaces coincide on a common axis.

Figure 4.2–27. This illustration clearly shows the radial and main thrust load carrying tapered surface of the cup race, the conical surfaces of the rollers and the tapered surface of the cone. A small thrust component is always carried by the backface cone rib. This is shown by the shaded portion of the end of the roller in contact with the cone rib. This contact, under load, aligns the rollers properly, preventing them from skewing. (Courtesy of the Timken Roller Bearing Company)

STEEP ANGLE TYPE TAPERED ROLLER BEARINGS. These bearings are designed to carry heavy thrust loads and are also well adapted to machine tools. (Figure 4.2–29)

THRUST TYPE TAPERED ROLLER BEARINGS. These bearings are de-

Prototype Machine

Figure 4.2–28. Standard type tapered roller bearing. A tapered roller bearing designed so that the radial capacity is about the same as or considerably higher than the thrust capacity is considered a standard bearing. (Courtesy of The Timken Roller Bearing Company)

Figure 4.2–29. Steep angle type tapered roller bearing. A tapered roller bearing designed so that the thrust capacity is much higher than the radial capacity is considered a steep angle bearing. (Courtesy of The Timken Roller Bearing Company)

4.3 Mechanical Drive Components

Figure 4.2–30. Thrust type tapered roller bearing. (Courtesy of Timken Roller Bearing Company)

signed to carry heavy thrust loads only. (Figure 4.2–30) Other types of bearings are used for centering, so it must be decided, considering the complete design, if other types of bearings which will carry combined radial and thrust loads can be used more profitably.

Preloading of Tapered Roller Bearings

To obtain maximum rigidity and minimum runout for machine tool spindles, it is desirable to preload the bearings. This is preferably done by the machine tool builder at the time of assembly. If a high production rate is prevailing, the machine tool builder, as a rule, sets up the preload requirements and keeps this record on an inspection sheet to get a uniform torque reading for all machines produced. It is always the best policy to work these requirements out in cooperation with the bearing manufacturer.

4.3 Mechanical Drive Components in Machine Tool Design

SPEED AND FEED RATIOS IN MACHINE TOOLS. Most speed and feed changes in machine tools are designed for belt or gear drives. It is well to remember that the best results are obtained when successive changes are designed for geometric progression, e.g., the lower speed or feed value is multiplied by the desired speed or feed ratio factor (r) to obtain the next successive higher speed or feed value.

Prototype Machine

SPEED RATIOS. It has been found by experience that nothing is gained if the speed ratio runs below 1.2, and the most practical ratio for ordinary machine tools should not be less than 1.3. For high grade machine tools, it is also best to keep the ratio under 1.5, and for machine tools of minor importance the ratio should not exceed 1.7.

FEED RATIOS. For high grade machine tools it has been found that the ratio between successive feeds should be 1.2 or less. If you have the minimum and maximum values and the number of intermediate values, and wish to check the ratio, the following equation may be used:

$$r = \sqrt[n-1]{\frac{b}{a}}$$

r = Ratio
a = The lowest speed or feed value
b = The highest speed or feed value
n = The total number of speed or feed values

The parts considered in this section are of decisive importance in

Figure 4.3–1. Flexible coupling which has torsional, axial, angular and parallel flexibility. (Courtesy of Dodge Manufacturing Corporation, Division of Reliance Electric Company)

4.3 Mechanical Drive Components

machine tool design. These parts are almost without exception purchased parts and the majority may be selected as standard units. Whenever possible, therefore, standard parts should be selected to keep the cost down. If special requirements are necessary, a firm commitment by the vender early in the design is the most desirable.

The mechanical drive components may be divided into the following main categories:

1. Couplings, Universal Joints, Clutches and Brakes
2. Chains and Sprockets
3. Belts, Pulleys and Sheaves
4. Gears, Ratchets, Geneva Motions, Linkages and Cams
5. Flexible Shafts

Couplings, Universal Joints, Clutches and Brakes

Two or more of these parts agree at times in their function. A clutch may do the work of a coupling or a brake and a universal joint is a variation of a coupling.

COUPLINGS. There are basically two main categories of couplings:

1. Flexible Couplings
2. Fixed Couplings

FLEXIBLE COUPLINGS. These couplings, in machine tools, are used principally to connect the shaft of a replaceable component, as an electric motor, with the drive of the machine. As a rule, the height alignment of an electric motor is controlled with a shim at final assembly of the machine tool, so if it becomes necessary to replace the motor, the manufacturing accuracy of the electric motor would determine the accuracy of the vertical alignment.

Horizontally, the alignment depends on the skill of the mechanic making the replacement. There are many types of couplings on the market that will satisfy this condition. The important thing to consider is to select a coupling that will satisfy one or more of the conditions mentioned below, depending on the requirements. (Figure 4.3-1)

1. Horsepower and Speed
2. Torsional Flexibility
3. Axial Flexibility
4. Angular Flexibility
5. Parallel Flexibility

Prototype Machine

Always select a coupling of adequate size to satisfy the horsepower of the motor used, even if the motor horsepower is more than the maximum requirements of the machine. Also make sure that the speed of the shaft has been considered, since coupling horsepower goes down with speed.

Torsional flexibility in most cases is advantageous because it will reduce shock. There may, however, be cases where it would be detrimental, as for an abnormally heavy pulsating load condition, or also if definite angular relationship must be maintained between the two shafts at all speeds and loads.

Axial flexibility in most cases is tolerable because the shafts under motion usually are restrained from axial motion by the bearings. Axial flexibility then will take care of expansion.

Angular and parallel flexibilities are permissible when it is just a matter of transmitting motion. In precision parts for machine tools, however, these characteristics are not desirable because connections are secured with a high degree of accuracy for the mating parts.

A high degree precision coupling is shown in Figure 4.3–2.

Figure 4.3–2. Semi-Universal CURVIC* coupling. (*CURVIC is a registered trademark of The Gleason Works). This coupling is manufactured with standard bevel gear machines. It has a zero degree pressure angle, which provides for axial freedom and constant backlash. A misalignment of two degrees maximum is possible. (Courtesy of Gleason Works)

4.3 Mechanical Drive Components

FIXED COUPLINGS. The common types of fixed couplings, also known as rigid couplings, are compression, ribbed and flanged.

COMPRESSION COUPLINGS. These couplings are so called because a thin slotted sleeve of the same inside diameter as the shafts, and slightly tapered on the outside diameter from each end coming to a high point in the middle, is compressed by bolting two flanged parts encompassing the tapered sleeve. There is, therefore, no need for a key in the shafts. These couplings are only good for light to medium loads.

RIBBED COUPLINGS. These couplings usually consist of halves, which when bolted together clamp the two shaft ends. There is no need for keys in the shaft, and they are only suitable for light to medium loads. They should only be used for machine tools where extreme accuracy is not required.

FLANGED COUPLINGS. These couplings may also be purchased. There are cases, however, where it is advantageous to make them in your own plant, as when they are part of another component. To facilitate timing, one half may have three holes in the flange for three clamp bolts, the other half may have a multiple of tapped holes, say twelve, or as many as the flange will permit.

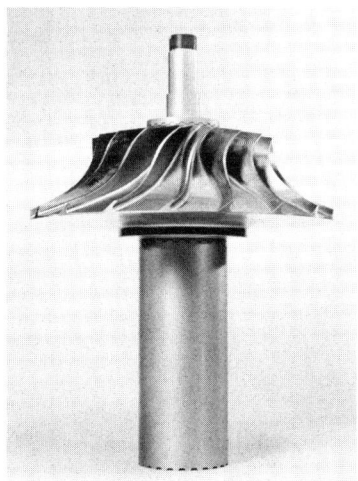

Figure 4.3-3. Fixed CURVIC* coupling. (*CURVIC is a registered trademark of the Gleason Works). This coupling is manufactured to the same high degree of precision as spiral bevel gears. Straight sided teeth, no profile curvature, and a pressure angle make it ideal for permanently fastening two or more parts together. This coupling provides for perfect centering, alignment and positive driving. (Courtesy of Gleason Works)

Prototype Machine

A high degree precision coupling for accurately connecting two precision shafts, or a precision part to a shaft, is shown in Figure 4.3–3. These couplings are manufactured by the same machines used for manufacturing spiral bevel gears, and, therefore, can be finished to the same high degree of accuracy as spiral bevel gears. The couplings do the centering, alignment and driving very accurately and, as a result, it is not necessary to have bearings close to the couplings except where loads are concentrated.

SPLINES. Common spline couplings are shown in Figure 4.3–4 and 4.3–5. These arrangements may be used for coupling important parts, such as gears, to a shaft.

A common coupling arrangement for a telescoping shaft, is shown in Figure 4.3–6. This design may have a large amount of axial freedom. Detailed description of many of these arrangements may be found in various engineering handbooks.

UNIVERSAL JOINTS. There are several types of these shaft connections on the market. The simplest and most common is the non-uniform velocity type. The uniform velocity types are more expensive. They

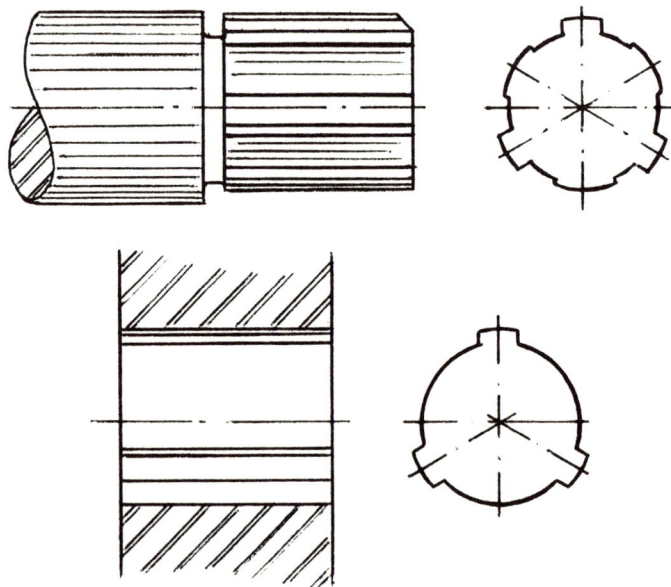

Figure 4.3–4a and Figure 4.3–4b. A three-spline coupling arrangement suitable for permanent fastening or replaceable fastening for heavy loads and higher speeds. This is a balanced design.

4.3 Mechanical Drive Components

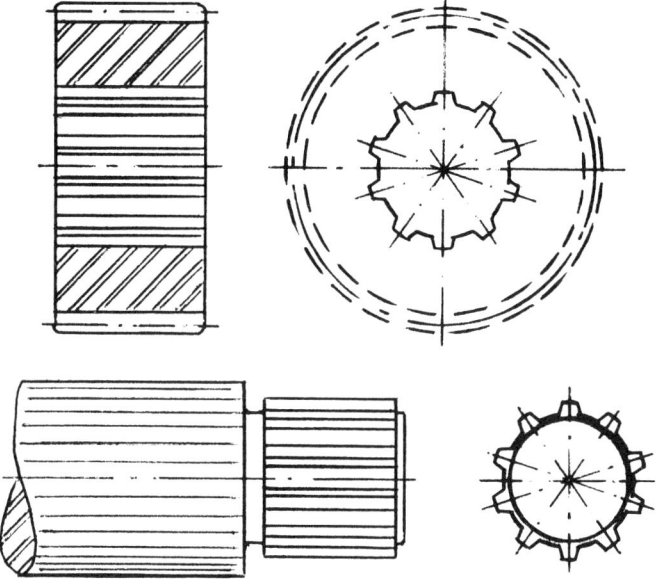

Figure 4.3–5a and Figure 4.3–5b. A ten-spline high precision coupling arrangement which requires a small radial space for heavy loads and higher speeds. This is a balanced design.

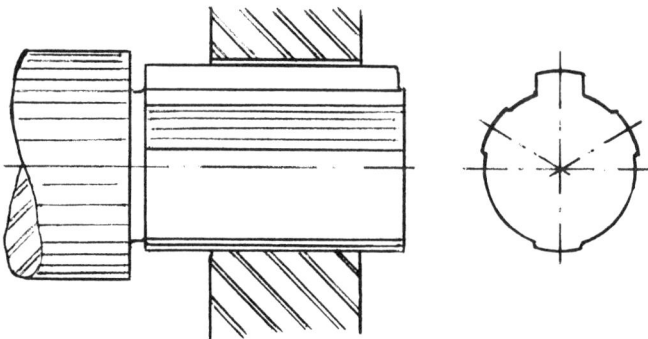

Figure 4.3–6. This is a solid single key coupling arrangement suitable for a large amount of axial freedom. Since this is not an exactly balanced design, it may cause some problems for extremely high speeds, when vibration cannot be tolerated.

will not be described in this book because, in most cases, a more dependable, less expensive design may be worked out better with properly arranged gear sets.

CLUTCHES. The most common types for machine tools are fine

Prototype Machine

tooth clutches, used mainly for maintaining a timed relationship in a drive. There may be several of these clutches in a machine drive because it may not be practical or economical to keep a fixed relationship between gear teeth and keys, or splines, in the process of manufacturing where a number of gears are required.

There may be many other reasons why a fine tooth clutch would be desirable. However, there are manufacturing limits to the ordinary fine tooth clutch, for the teeth cannot be made too fine and shallow. The tooth depth for a clutch having 180 teeth and a 2¾ inch outside diameter, for instance, would be about .040 inch, which would be considered close to the practical limit. This would provide for ½ degree adjustment.

If this is not fine enough, a differential clutch may be the best solution. This type of a clutch can be designed to meet almost any requirements. Figure 4.3-7 shows one such design. The simplest arrangement would be to have the clutch adjustable to the conventional units of measurements, degrees and minutes. Therefore, we may set up the following formula:

$$\frac{360 \times 60}{N \times n} = 1, \text{ where } 360 \times 60 = \text{number of minutes in one turn of the shaft}$$

N = number of teeth in one clutch face
n = number of teeth in other clutch face
1 = finest increment of adjustment = one minute

$$\frac{360 \times 60}{N \times n} = \frac{2 \times 2 \times 2 \times 2 \times 2 \times 3 \times 3 \times 3 \times 5 \times 5}{\underbrace{2 \times 2 \times 3 \times 3 \times 5}_{N} \times \underbrace{2 \times 2 \times 2 \times 3 \times 5}_{n}}$$

$N = 180$
$n = 120$

If 10 minutes is the smallest increment of adjustment, the formula would read:

$$\frac{360 \times 60}{N \times n} = 10, \text{ or } \frac{360 \times 6}{N \times n} = \frac{2 \times 2 \times 2 \times 2 \times 3 \times 3 \times 3 \times 5}{\underbrace{2 \times 2 \times 3 \times 5}_{N} \times \underbrace{2 \times 2 \times 3 \times 3}_{n}}$$

$N = 60$
$n = 36$

4.3 Mechanical Drive Components

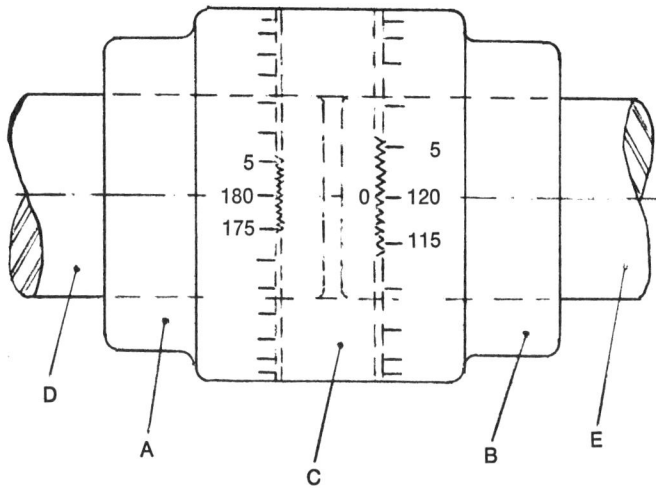

Figure 4.3-7. Differential clutch. A, 180 teeth clutch member, is keyed to shaft D and slides out of engagement for adjustment. B, 120 teeth clutch member, is keyed to shaft E and slides out of engagement for adjustment. C, central clutch member with 180 teeth on one face and 120 teeth on the other face, is free to turn on shaft D and shaft E. Clutch members A and B are graduated every fifth tooth for easy setting.

Figure 4.3-8. CURVIC* clutch type coupling. All CURVIC* couplings, being manufactured on spiral bevel gear machines may be produced with localized tooth bearing. In most cases, convex teeth are used on one member and concave teeth on the other member. (Courtesy of Gleason Works). *CURVIC is a registered trademark of the Gleason Works.

This arrangement is also applicable to rectilinear clutch design or a differential screw design, which will give a fine linear adjustment.

Prototype Machine

Other types of clutches that are useful when no timed relationship is necessary are square jaw, spiral jaw or friction clutches.

An accurate clutch designed and manufactured to the same high degree of accuracy as spiral bevel gears is shown in Figure 4.3–8.

BRAKES. For machine tools, a brake is generally applied at the motor. There may, however, also be other places where a brake is required.

Chains and Sprockets

This drive is positive and finds many types of applications where the center distance is too short for a belt drive, or too long for a set of gears. (Figure 4.3–9)

However, when a high degree of accuracy is required, or where vibration and noise cannot be tolerated, the design should be changed to accommodate a gear drive.

Speed is, however, no problem for a good chain drive, and 10,000 rpm is not unusual.

Belts, Pulleys and Sheaves

An endless belt is recommended for a grinding wheel drive and similar equipment when a high degree of accuracy is required and noise or vibration cannot be tolerated. (Figures 4.3–10, 4.3–11 and 4.3–12)

Figure 4.3–9. Chain and sprocket drive. (Courtesy of Dodge Manufacturing Corporation, Division of Reliance Electric Company)

4.3 Mechanical Drive Components

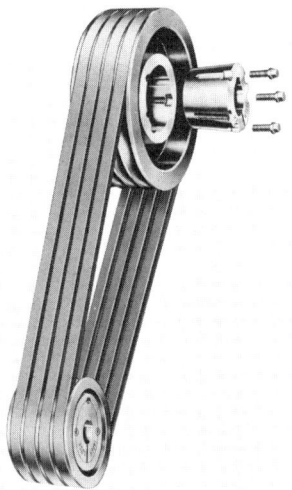

Figure 4.3–10. Endless V-belt drive. (Courtesy of Dodge Manufacturing Corporation, Division of Reliance Electric Company)

Figure 4.3–11. Sheave for V-belt drive. (Courtesy of Dodge Manufacturing Corporation, Division of Reliance Electric Company)

Gears, Ratchets, Geneva Motions, Linkages and Cams

GEARS. One of the most important components in a power drive is a set of gears. They can be arranged in a great number of varieties and combinations to satisfy many problems in machine tool design.

Epicyclic or planetary gearing can be arranged for a large reduc-

Prototype Machine

Figure 4.3–12. Timing belt drive. This is a positive drive arrangement, providing a quiet drive with no slippage. (Courtesy of Dodge Manufacturing Corporation, Division of Reliance Electric Company)

Figure 4.3–13. Epicyclic gearing indicating the linear progress of a .100 inch pitch adjusting screw to within .001 inch increments, keeping accurate record up to 19.2 inches with small circular dials.

The ratio may be calculated from the following equation:

$$R = 1 - \frac{A}{p} \times \frac{P}{B} \qquad (1)$$

A = Fixed internal gear containing zero marks for .001 inch indication, and linear progress of adjusting screw in increments of .100 inch up to a maximum of 19.2 inches.
B = Driven internal gear.
C = Center of gears A and B and center of adjusting screw. Also center of driver bracket D.
D = Driver bracket driving planet pinions p and P. This bracket also contains graduation dial indicating linear progress of adjusting screw in increments of .001 inch. This dial has 100 graduations, and one turn of the dial indicates .100 inch, which is the pitch of the screw.
p = Planet pinion engaging fixed internal gear A.
P = Planet pinion engaging driven internal gear B.
R = Rotation of driven gear B per revolution of driver bracket D.

$$\text{One revolution of driven gear B} = \frac{1}{R} \qquad (2)$$

4.3 Mechanical Drive Components

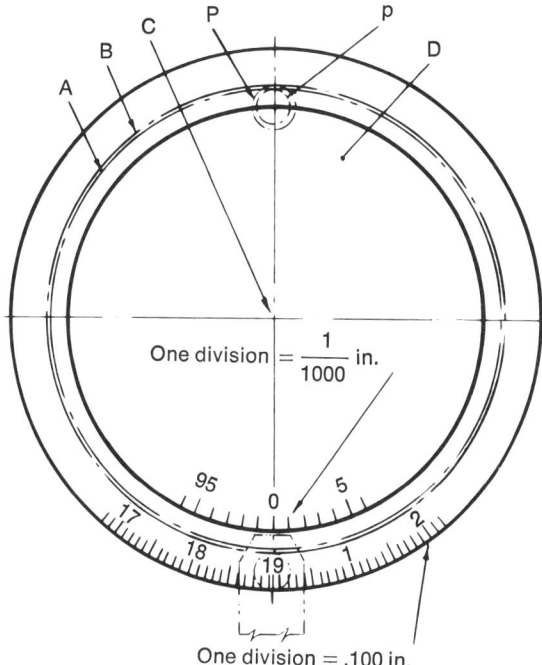

Assume:
A = 192 teeth
B = 193 teeth
p = 16 teeth
P = 16 teeth
Diametral pitch = 64
A pitch diameter = 192/64 = 3.000 inches
B pitch diameter = 193/64 = 3.0156 inches
p and P pitch diameters = 16/64 = .250 inch

For indicating an adjustment, the gears can be modified to accept the differences in gears A and B, still showing accurate results.

From equation 1, we have:

$$R = 1 - \frac{192}{16} \times \frac{16}{193} = 1 - \frac{192}{193} = 1 - .9948 = .0052$$

From equation 2, we have:

$$\frac{1}{R} = \text{One revolution of driven gear B} = \frac{1}{.0052} =$$
192.2 revolutions of driver bracket D.

Prototype Machine

tion in a very small space, which is often very valuable for fine adjustments. Sometimes, this gearing is used to indicate the progress of an adjusted member to a close degree of accuracy on a couple of small circular disks. (See figure 4.3–13)

A differential gear set in a train of gears can be arranged to vary the speed of one portion of the drive relative to another portion by applying a cam motion to the differential housing. (See figures 4.3–14a and 4.3–14b)

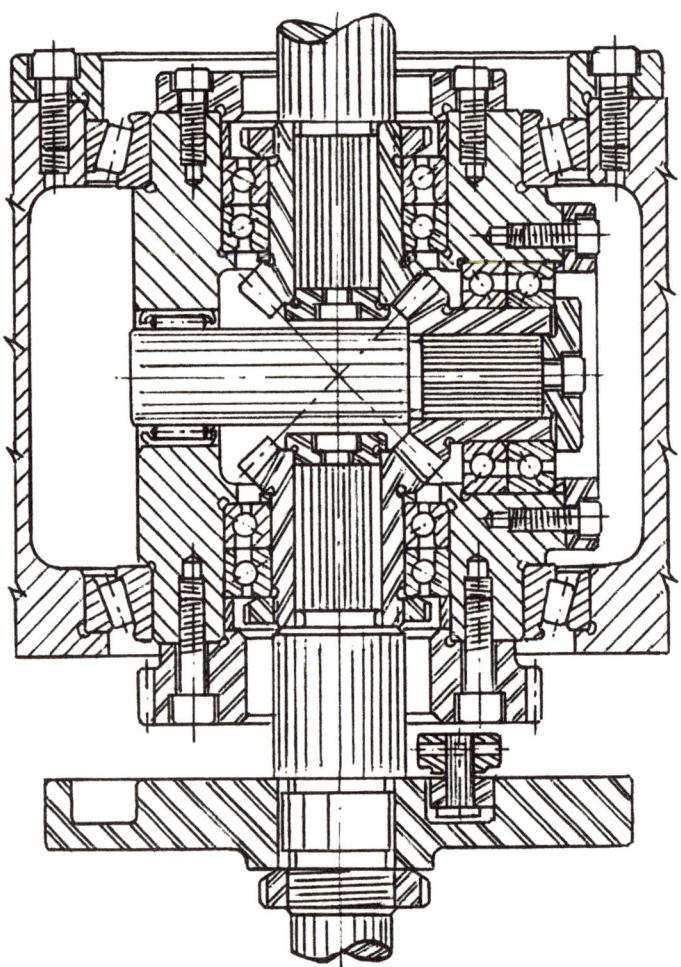

Figure 4.3–14a. Differential gear set. This arrangement of gears, in connection with other drive components, may be used to vary a portion of the drive gradually, intermittently or to many desired patterns while the drive is running.

4.3 Mechanical Drive Components

A differential gear set can also be arranged to advance or retard in steps one portion of the drive relative to another portion by connecting a geneva mechanism (Figure 4.3–14c) to the differential housing, or by having a cam in connection with a ratchet wheel advance the differential housing. (Figure 4.3–14d)

Gears for machine tools can be divided into the following main classifications:

1. Spur
2. Worm
3. Bevel

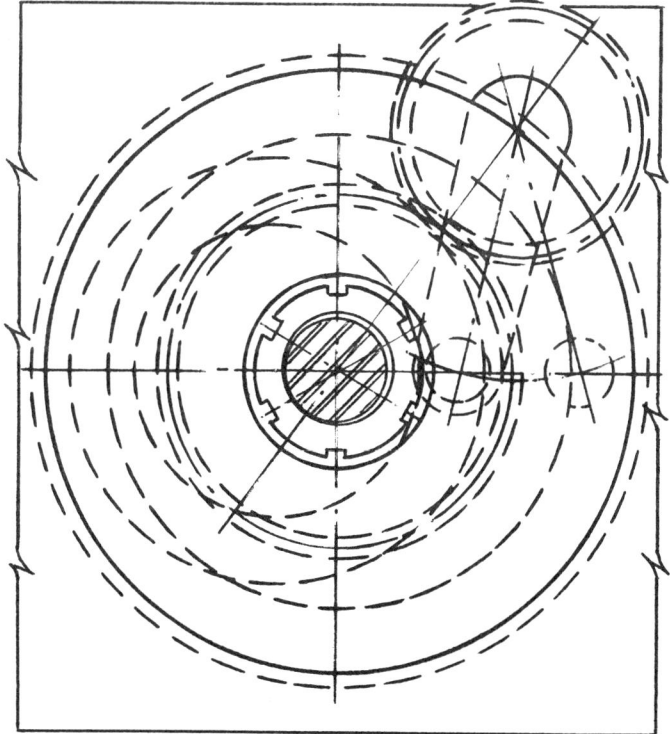

Figure 4.3–14b. Cam and spur gear control for differential gear set. Differential gear set shown in figure 4.3–14a, as designed, may be operated by a cam and a pair of spur gears. The driven spur gear is centrally fastened to the differential housing. The driver gear is fastened and centered to the fulcrum end of the lever, which turns on a stud, fastened to the fixed part of the machine. The end of the lever carries a roller operated by the cam track. The cam is fastened to the driver shaft of the differential.

It is thus possible to vary the speed of the driven differential shaft in relation to the driver shaft. Both the cam and the spur gears can be used for this variation.

Prototype Machine

Spur Gears

These gears are extensively used in machine tool drives to transmit motion for parallel shafts. The pitch diameters can be kept to close tolerances, so no radial adjustments are necessary at assembly. The following types are the most common for machine tools:

1. Plain Straight Tooth Spur Gears and Racks

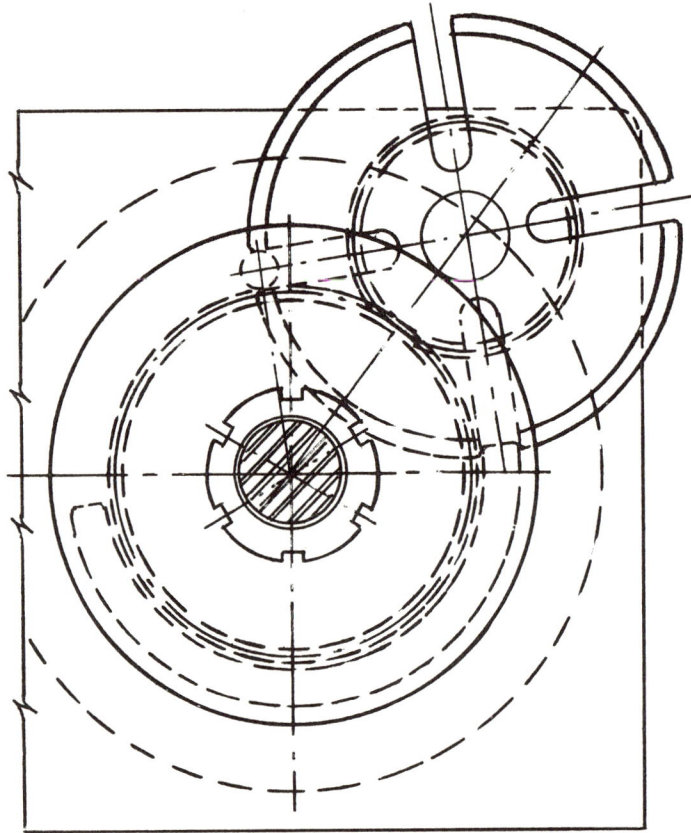

Figure 4.3–14c. Geneva control for differential gear set. Differential gear set shown in Figure 4.3–14a, may also be designed for intermittent advancement of the driven differential shaft in relation to the driver shaft. The geneva driver plate is fastened to the driver shaft in place of the cam. The driven geneva wheel is fastened to the driver spur gear, and the driven spur gear is fastened to the differential housing as shown in Figure 4.3–14a.

The differential housing is locked with the geneva drive plate between advancements. This lockup is explained in Figure 3.9–2.

If the reversed action is desired in the driven shaft, an idler gear may be placed between the two spur gears.

4.3 Mechanical Drive Components

2. Helical Tooth Gears and Racks
3. Herringbone and Double Helical Gears

PLAIN STRAIGHT TOOTH SPUR GEARS AND RACKS. These gears are the most common, having teeth parallel to the axis or 90 degrees to the rack sides. They are the least expensive and satisfy a great number of applications.

HELICAL TOOTH GEARS AND RACKS. These gears are modifications of the standard straight tooth spur gear. They are excellent for parallel shaft applications, and have a greater load carrying capacity than the

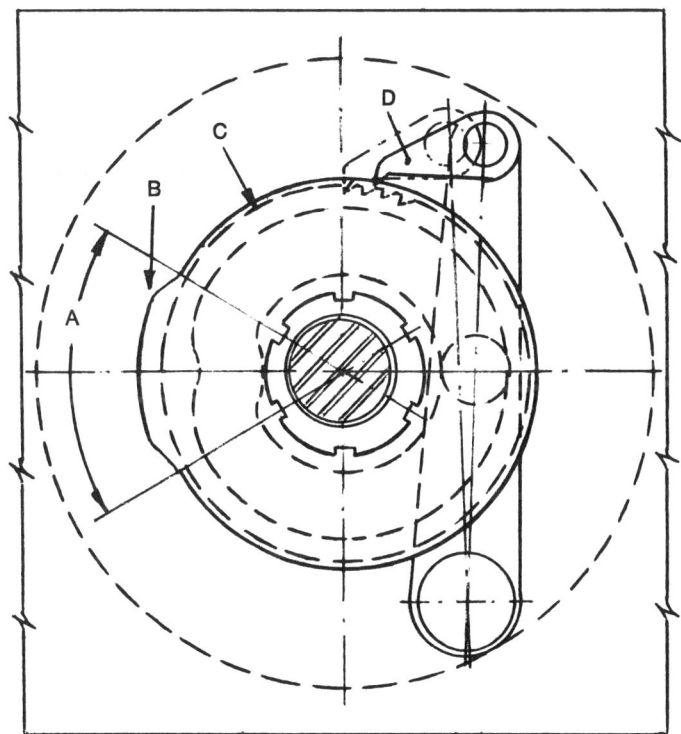

Figure 4.3–14d. Cam and ratchet wheel control for differential gear set. This is a simple arrangement for intermittent advancement of the driven differential shaft in relation to the driver shaft. The cam is fastened to the driver shaft as shown in Figure 4.3–14a. The index portion A of the cam track moves the roller toward the center of the drive shaft, moving the ratchet pawl D toward the left. The ratchet pawl may be spring engaged. The cam may be designed to turn the ratchet wheel one or more teeth at a time. The outside of the cam may have a cam lobe B for disengaging a lockup pawl or a brake load, if necessary. The ratchet wheel C is fastened to the differential housing in place of the spur gear shown in Figure 4.3–14a.

Prototype Machine

equivalent straight tooth spur gear. They will also operate satisfactorily at much higher speeds. As a result of the helix angle of the teeth to the axis, the teeth overlap and the motion is therefore transmitted more smoothly and quietly than with straight tooth spur gears. The helix angle induces a thrust load on the shaft. This is, however, no problem with antifriction bearings. (Figure 4.3-15)

HERRINGBONE AND DOUBLE HELICAL GEARS. These gears are also a modification of the straight tooth spur gear. As a result of the opposing helix angles, no thrust load is set up in the shaft. They are used for heavy loads and high speeds.

CROSSED AXIS HELICAL GEARS. These gears are also known as spiral gears. They cannot be recommended for machine tool applications because they have point contact only between the mating teeth. Since considerable sliding takes place, excessive wear is a problem.

Worm Gears

This is one of the oldest types of gear drives, and is very useful in large speed reduction through one stage. For good accurate results, a hardened and ground steel worm with a cast iron worm gear may be used. A well lubricated combination may be run up to 700 feet per minute, depending on the load.

Up to 1000 feet per minute is possible with moderate load for a hardened and ground worm and a good grade bronze worm gear well lubricated. The oil should be applied at the worm for the higher speeds.

Some basic considerations should be observed for machine tool drives.

1. For power drives, in particular, the helix angle of the worm should never be less than 9 degrees, and preferably over 12 degrees. It should never be over 30 degrees, since nothing is gained in performance. This, for all practical purposes, would call for a worm of three or four threads. The sum of the threads in the worm and the number of teeth in the worm gear should never be less than 37 for good tooth action.
2. For drives where a high reduction through a single stage is the primary purpose, a single or double threaded worm may be considered, with a consequently lower helix angle and a lower degree of efficiency. One, two, three and four threads for the

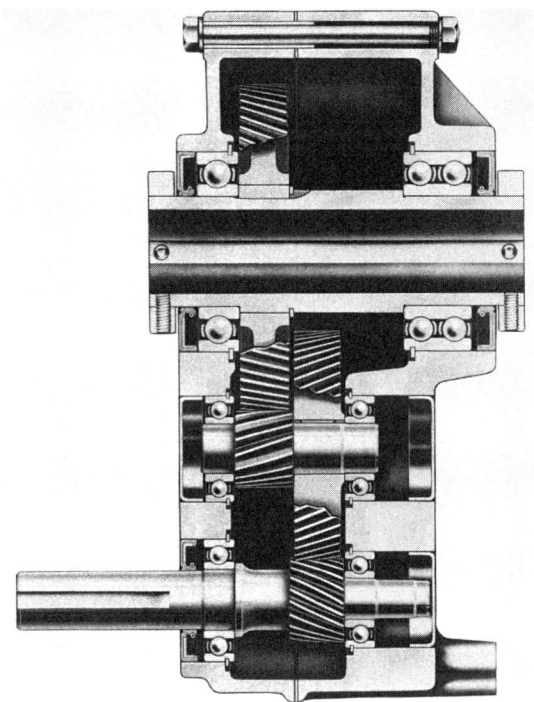

Figure 4.3–15. This shows a good application of parallel shaft, helical tooth gears. (Courtesy of Dodge Manufacturing Corporation, Division of Reliance Electric Company)

worm will usually cover most requirements for machine tool design. If possible, use the hand of thread (right or left) for which a hob is available, because the direction of rotation, as a rule, can be controlled with other gear sets in the drive.

Bevel Gears

With the efficient, accurate and comparatively inexpensive methods now available for manufacturing bevel gears, together with good antifriction bearings now on the market, these gears are the best drive components for transmitting power with nonparallel transmission axis.

Bevel gears may be classified in the following main groups:

1. Straight Bevel and ZEROL * Bevel
2. Spiral Bevel
3. Hypoid
4. High Reduction Hypoid

*ZEROL is a registered trademark of The Gleason Works

Prototype Machine

Figure 4.3–16. A pair of straight bevel gears with localized tooth bearing. This design will permit a slight misalignment at assembly, and a small deflection under load will not concentrate the load at the heel of the tooth. (Courtesy of Gleason Works)

STRAIGHT BEVEL GEARS. These bevel gears are the least expensive to manufacture, and are extensively used in machine tools for auxiliary equipment as well as power drives. They are well suited for speeds up to 1000 feet per minute. Straight bevel gears produced on the latest bevel gear machines have localized tooth bearing. This is a definite advantage in that a slight misalignment at assembly or a slight deflection under load will not concentrate the load at the heel of the teeth. (Figure 4.3–16)

ZEROL bevel gears have the same axial loads as the equivalent straight bevel gear set. They may be ground if hardened gears are required. (Figure 4.3–17)

SPIRAL BEVEL GEARS. These gears have a continuous pitchline contact due to the spiral angle of the teeth, and therefore, will operate more smoothly and quietly than straight bevel or ZEROL bevel gears. When peripheral speeds in excess of 8000 feet per minute are encountered, ground pairs should be used. These gears are produced with localized tooth bearing. (Figure 4.3–18)

HYPOID GEARS. These gears are often used to great advantage in machine tools. One of the advantages in a pair of hypoid gears is that

4.3 Mechanical Drive Components

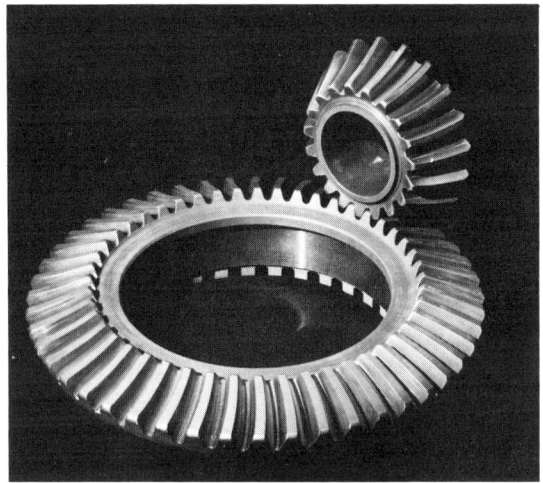

Figure 4.3-17. ZEROL* bevel gear pair. (Courtesy of Gleason Works). *ZEROL is a registered trademark of The Gleason Works.

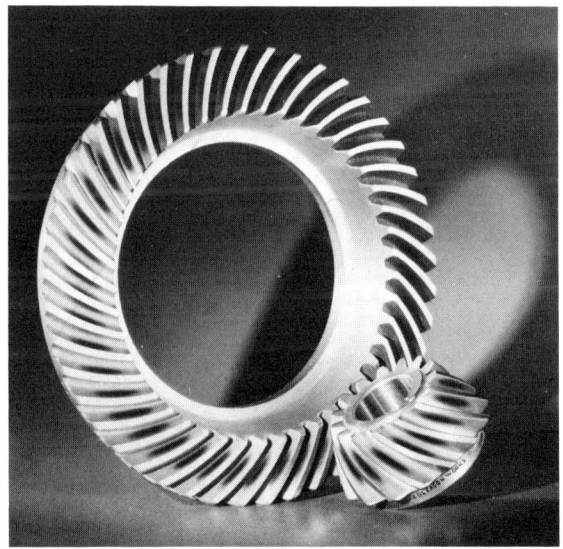

Figure 4.3-18. Spiral bevel gears. (Courtesy of Gleason Works)

the pinion is of a larger diameter than the equivalent spiral pair, thus permitting a larger ratio without making the pinion too small. Another advantage is that the pinion center is offset in relation to the gear center, making possible a crossed shaft design.

The conventional hypoid gears are made for ratios up to 10 to 1.

They are usually arranged and designed so that the pinion has a larger spiral angle than the gear. Therefore, there is a sliding action in the lengthwise direction of the teeth.

Extreme pressure lubricants are recommended for hypoid gears and spiral bevel gears subject to extreme conditions of shock, severe starting conditions, or heavy loads. (Figure 4.3–19)

HRH * (HIGH REDUCTION HYPOID) GEARS. These gears have found many applications in machine tool design. For ordinary applications in machine tools, the ratio is usually from 10 to 1 to approximately 120 to 1. They may also be designed for reductions as high as 360 to 1, and even greater reductions are possible if the pitch required for such a reduction is not too fine to be practical.

The pinion may be made cylindrical or conical. The advantages of a cylindrical pinion is the increased diameter in front, which permits the use of a rigid straddle mounting. Since both members in a pair may be made of case hardened steel and may be accurately ground, they are well suited for precision drives in machine tools, especially accurate index drives. (Figures 4.3–20 and 4.3–21)

Modern bevel gears may now be tested accurately to assure correct tooth bearing before being assembled in a machine tool. (Figure 4.3–22)

*HRH is a registered trademark of The Gleason Works

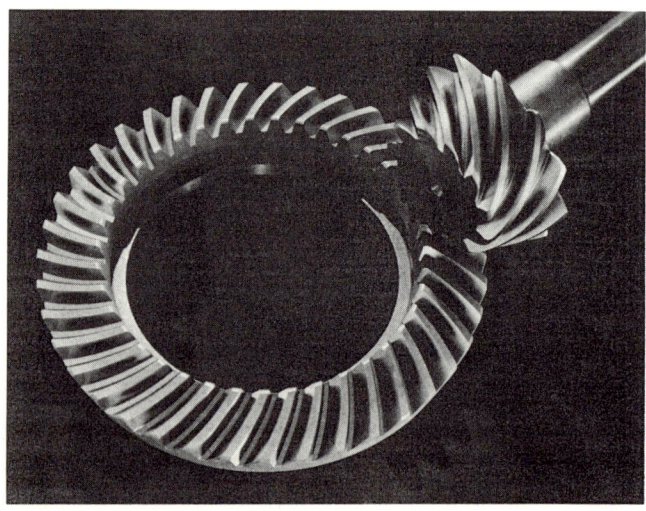

Figure 4.3–19. Hypoid gears. (Courtesy of Gleason Works)

4.3 Mechanical Drive Components

Figure 4.3-20. HRH* (High reduction hypoid) gear with cylindrical pinion. (Courtesy of Gleason Works). *HRH is a registered trademark of The Gleason Works.

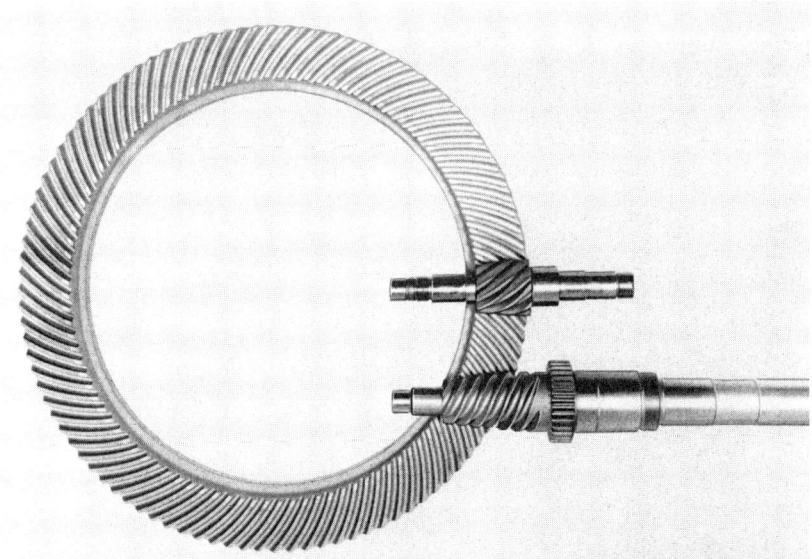

Figure 4.3-21. HRH* (High reduction hypoid) gear with two conical pinions. (Courtesy of Gleason Works). *HRH is a registered trademark of The Gleason Works.

Prototype Machine

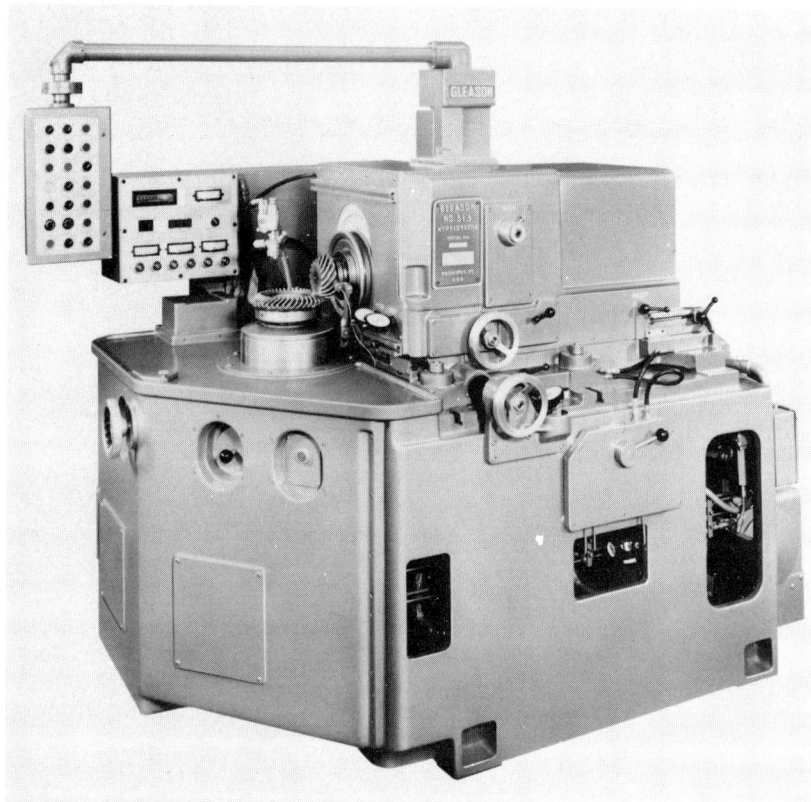

Figure 4.3–22. Bevel gear testing machine. This machine provides a practical method for determining that the gears are correct before being assembled, and for controlling precisely the quality of the gears during all stages of their manufacture.

This is important because correct tooth shapes, uniform spacing of the teeth, and teeth concentric with the bore or shank all are necessary requirements for smooth, quiet running gears.

The location of the tooth bearing can also be definitely determined, considering the loads expected. (Courtesy of Gleason Works)

Ratchets

One of the simplest methods to produce intermittent rectilinear or rotary motion is the use of ratchet mechanism. A simple ratchet wheel is seen in Figure 4.3–14d.

The ratchet wheel may be mounted so that it is interchangeable from a wheel of few tooth numbers to one with many; or it may be a fine tooth ratchet wheel operated by an adjustable pawl to give a

4.3 Mechanical Drive Components

variety of stroke lengths, advancing the wheel one tooth per stroke or several teeth as the case may be. The pawl may be operated mechanically by a cam or if air or hydraulic power is available, it may be operated pneumatically, hydraulically or electrically by a solenoid or a rectilinear or rotary actuator. There are then many possibilities for locking the ratchet wheel between indexing:

1. Cam-released and spring-loaded lockup pawl engaging a slot in a disk fastened to the ratchet wheel.
2. Air-operated in both directions, or for safety, air-released and spring-loaded.
3. Hydraulically operated in both directions, or for safety, hydraulically released and spring-loaded.
4. Electric actuator release and spring-loaded.
5. If a high degree of accuracy is not required for intermittent advancement, any type of brake may be sufficient for holding the ratchet wheel between indexing.

Geneva Motion

This type of intermittent motion as shown in Figure 3.9–2 offers a great range of indexing. It may be arranged for interchangeable index plates, or may be worked out in connection with a properly arranged gear train. (See figure 4.3–14c)

Linkages

Some of the most common linkages are shown in engineering handbooks.

Cams

This is an important mechanical drive component. There are several types of cams. Some of the main shapes encountered in machine design are:

1. Disk cam, Figure 4.3–23
2. Face cam, Figure 4.3–24
3. Cylinder or barrel cam, Figure 4.3–25

The general design of the machine tool determines the type of cam that is best suited for the machine, considering performance and ease of replacement.

Prototype Machine

The paths of these cams must be designed to produce certain desired motions. These motions are often governed by the requirements of functions of the machine, such as cutting or grinding, or moving a

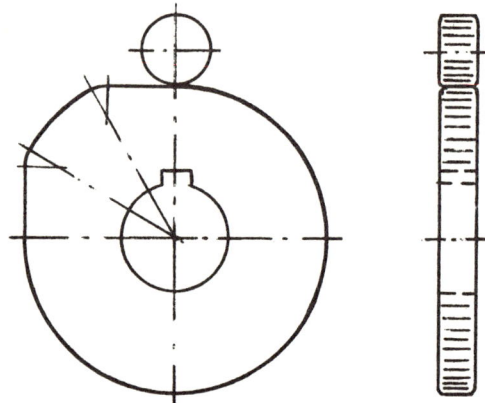

Figure 4.3–23. Disk cam. This cam may operate a roller or numerous variations of rider arrangements.

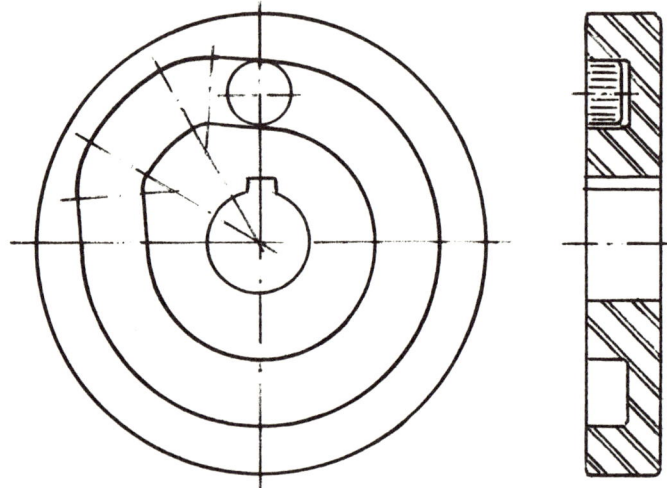

Figure 4.3–24. Face cam. This cam operates a roller. It is sometimes called a complementary cam, because the enclosed path completes the action without the help of springs or other components. There should always be a small clearance between roller and path, since rolling can occur only when one side of the roller diameter is in contact.

4.3 Mechanical Drive Components

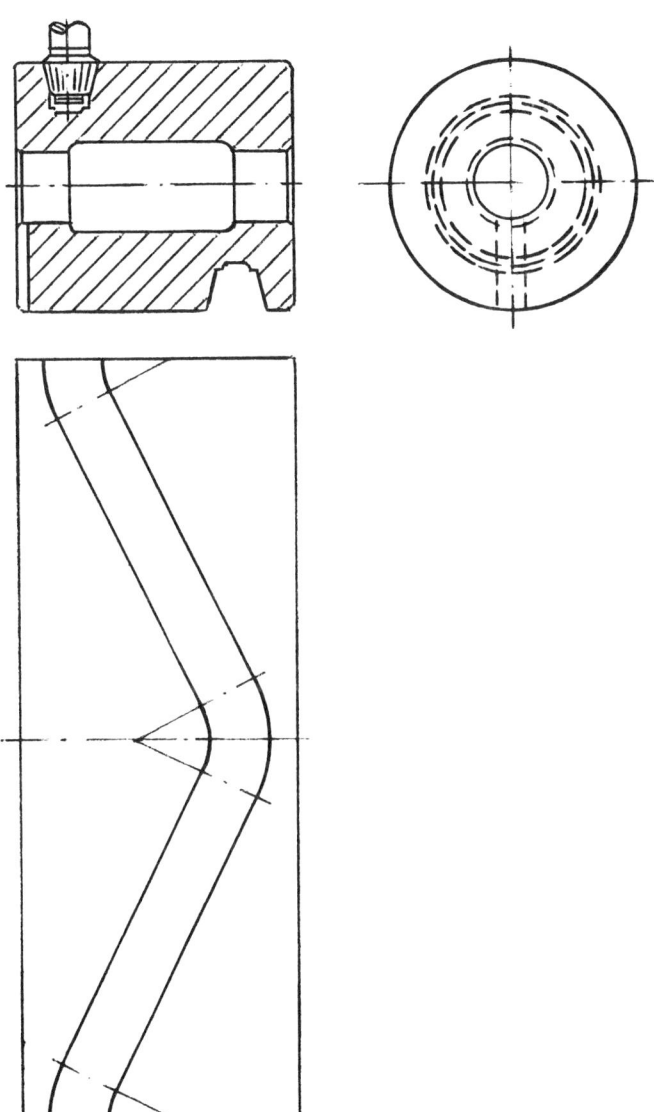

Figure 4.3–25. Cylinder cam, sometimes also called barrel cam. This cam operates a roller. The roller should be conical to approach a true rolling action. Since the spiral angle of the path is seldom if ever constant throughout its length, the angle of the roller must be made to suit the most important portion of the path, sometimes where the load is the heaviest.

Prototype Machine

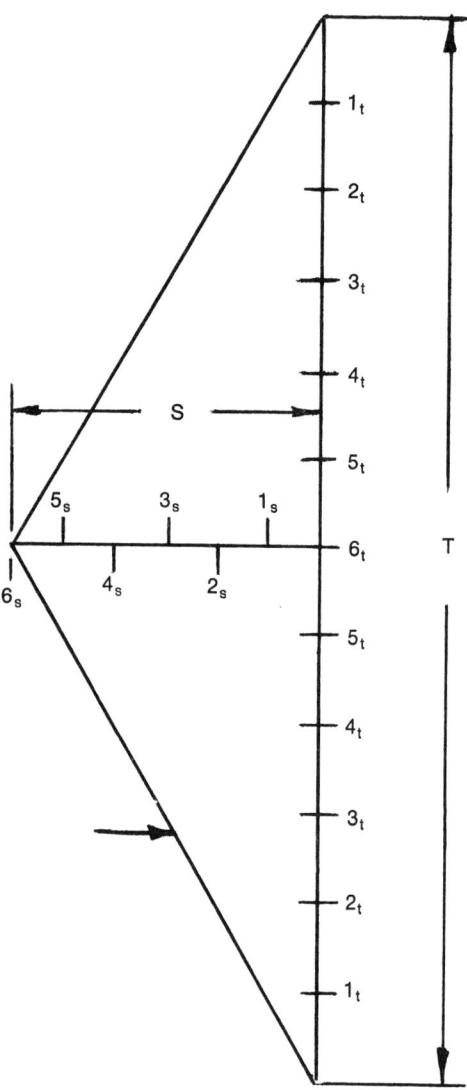

Figure 4.3–26a. Uniform motion cam path layout for rectilinear motion or cylinder cam. For cylinder cam, T is the circumference of the cam and S is the axial distance travelled by the follower. The cam path is always for the sharp point of a follower or the center of a roller. If the follower or cam rider has a radius or a roller is used, always modify the path to suit the radius of the follower or roller. If a swinging lever is used, the arc of the roller travel must also be considered.

4.3 Mechanical Drive Components

machine component along a desired path. There are three main shapes of cam paths for machine tools:

a. Uniform Motion, figure 4.3–26
b. Harmonic Motion, figure 4.3–27
c. Uniformly Accelerated and Retarded Motion, figure 4.3–28

A cam path may be designed for one of these shapes, or may contain two or all three, or modifications of them. The time available to perform a function, together with other requirements, will determine the shape of the entire path.

A uniform motion path, which will move a machine component through equal increments of space in equal increments of time, may be the most desirable shape to perform a certain function. This path

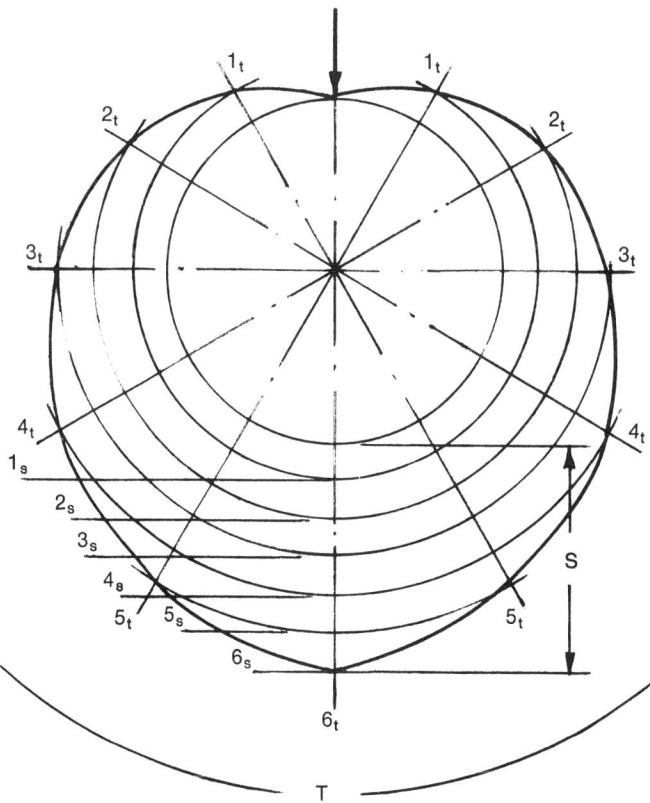

Figure 4.3–26b. Uniform motion cam path for disk cam or face cam. T in this case is the full circumference of the cam.

Prototype Machine

will, however, start and stop motion abruptly, so if speeds are high, it may be necessary to apply harmonic motion or uniformly accelerated motion at the beginning of advancement or reversal of motion to make the motion smooth.

There are many other shapes of cam paths which are determined

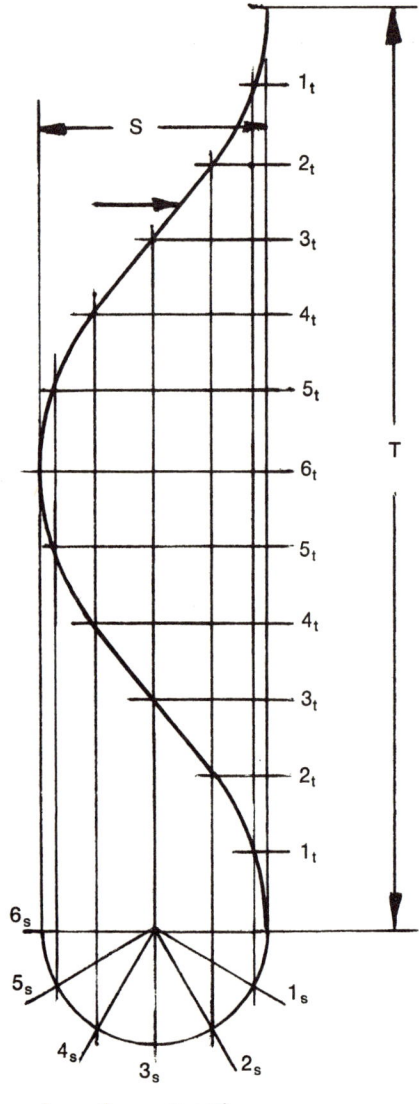

Figure 4.3–27. Harmonic motion cam path.

4.3 Mechanical Drive Components

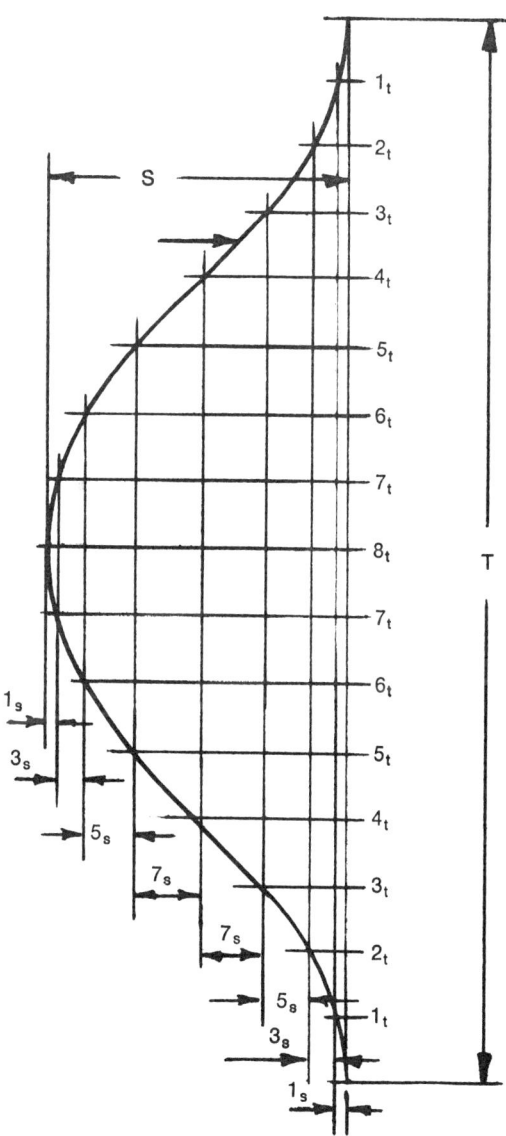

Figure 4.3-28. Uniformly accelerated and retarded motion cam path. Point follower or roller center in equal units of time, t, moves at uniformly increasing numbers of space units. Thus, in the first time unit the follower moves one space unit, in the second time unit the follower moves $1 + 2 = 3$ space units, in the third time unit the follower moves $3 + 2 = 5$ space units etc.

Prototype Machine

by the requirements of the machine. These shapes may not be like any of the standard shapes. As a rule, it is possible to include one of the standard shapes at the point of reversal to make the action smooth.

For high speed machinery it should be remembered that the manufacturing of the cam is of great importance. Undesirable effects are often the result of inaccurate manufacturing and imperfect material.

Flexible Shafts

These shafts find limited use in machine tools. They may be used for remote control of auxiliary equipment of minor importance.

4.4 Electrical Components in Machine Tool Design

Standardization of replaceable components for machine tools is a desirable characteristic. The aim of the design engineer should always

Table 4.4–1. Frame Assignments for Open and TEFC Continuous Duty 60cps Polyphase Squirrel Cage Motors *

HP	3600 rpm	1800 rpm	1200 rpm	900 rpm
½				182
¾			182	184
1		182	184	213
1½	182	184	184	213
2	184	184	213	215
3	184	213	215	254U
5	213	215	254U	256U
7½	215	254U	256U	284U
10	254U	256U	284U	286U
15	256U	284U	324U	326U
20	284U	286U	326U	364U
25	286U	324U	364U	365U
30	324S	326U	365U	404U

* Courtesy of General Electric Co.

4.4 Electrical Components

be interchangeable parts, especially when there is a chance that the parts may have to be replaced.

The electrical industry has done a lot in this respect. The National Electrical Manufacturers' Association (NEMA), has arranged for a rerate program, assigning new frame numbers for standard motors. This is shown in Table 4.4-1. The dimensions for foot mounted, face mounted and flange mounted a-c motors according to frame numbers are shown in Tables 4.4-2a, 4.4-2b, 4.4-2c and 4.4-2d.

The automotive industry has long been using many types of machine tools for producing a great variety of parts. Because the design engineer of these different machine tools often would not be familiar with conditions in automotive plants, the location of the electrical components would vary considerably. Therefore, it became necessary to establish electrical standards for industrial equipment for the machine tool manufacturers doing business with the automotive industry. These standards are especially helpful to the electrical engineer preparing the electrical work for the machine tool. Some suggestions in these standards are very helpful to the design engineer also, even for machine tools supplied outside the automotive industry. It is suggested in these standards that the location of electrical equipment be such that it will be easily accessible for service, and not be subject to damage by other equipment used around the machine. The standards were drawn up by the Joint Industry Conference (JIC), and the following three points are of interest to the design engineer:

1. Safety for personnel
2. Uninterrupted production
3. Long equipment life

Standard Motor Mountings Adopted by NEMA

Three main types of mountings for electric motors are used for main drive motors for machine tools or for the various accessory components:

1. Foot mounted motors, the most common for machine tool drives. (Figure 4.4-3)
2. Face mounted motors, which quite often come with feet. A brake or a gear reduction unit may be mounted to the face, with the screws entering the motor face and the driven unit or

Prototype Machine

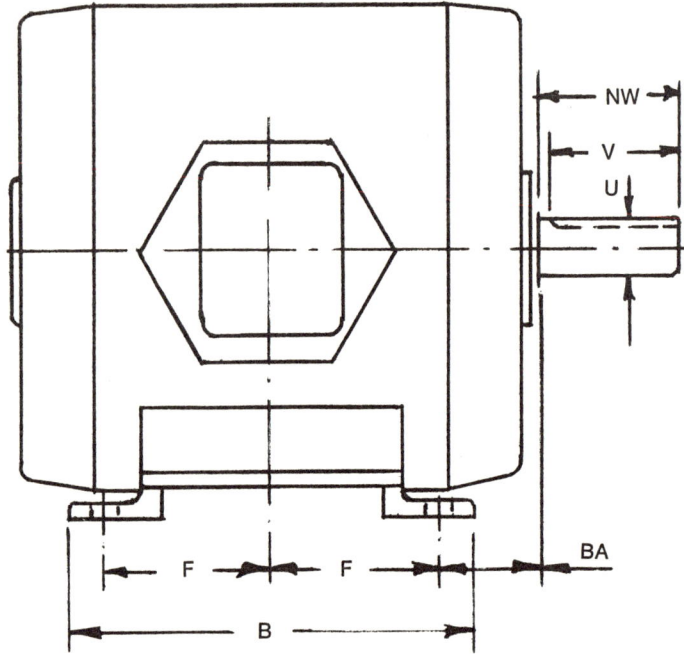

brake centered to the motor shaft by a male rabbet on the face of the motor. The motor may then be fastened by the feet, usually in a horizontal position. (Figure 4.4-4)

3. Flange mounted motors are sometimes used for drives on machine tools. The motor is fastened by screws through the motor flange and centered to the driven shaft with a male rabbet on the face of the motor. (Figure 4.4-5)

Standard motors are manufactured with various features, designed to suit the requirements of a particular machine tool. Such designs are:

1. Drip-proof motors. These are open motors, where the ventilating openings are so designed that liquid or solid objects striking the motor in a near vertical direction will not interfere with the operation of the motor. These motors are used for many machine tool applications.
2. Totally enclosed nonventilated motors (TENV). These motors also come with the enclosures designed to withstand explosion and are then called explosion-proof motors.

4.4 Electrical Components

Table 4.4–2a. Dimensions for Foot Mounted a-c Motors *

Frame Number	B	BA	F	NW	U	V Min
182	6½	2¾	2¼	2¼	⅞	2
184	7½	2¾	2¾	2¼	⅞	2
213	7½	3½	2¾	3	1⅛	2¾
215	9	3½	3½	3	1⅛	2¾
254U	10¾	4¼	4⅛	3¾	1⅜	3½
256U	12½	4¼	5	3¾	1⅜	3½
284U	12½	4¾	4¾	4⅞	1⅝	4⅝
286U	14	4¾	5½	4⅞	1⅝	4⅝
324U	14	5¼	5¼	5⅝	1⅞	5⅜
324S	14	5¼	5¼	3¼	1⅝	3
326U	15½	5¼	6	5⅝	1⅞	5⅜
326S	15½	5¼	6	3¼	1⅝	3
364U	15¼	5⅞	5⅝	6⅜	2⅛	6⅛
364S	15¼	5⅞	5⅝	3¾	1⅞	3½
365U	16¼	5⅞	6⅛	6⅜	2⅛	6⅛
365S	16¼	5⅞	6⅛	3¾	1⅞	3½
404U	16¼	6⅝	6⅛	7⅛	2⅜	6⅞

* Courtesy of General Electric Co.

3. Totally enclosed fan-cooled motors (TEFC). They are usually equipped with an external fan and fins to dissipate the heat. This motor finds many uses in machine tools.

The standard polyphase squirrel cage motors are the most common, and may be grouped as follows according to electrical characteristics:

1. Integral horsepower a-c induction motor. The most common for machine tools is the constant speed motor. The speed varies

Prototype Machine

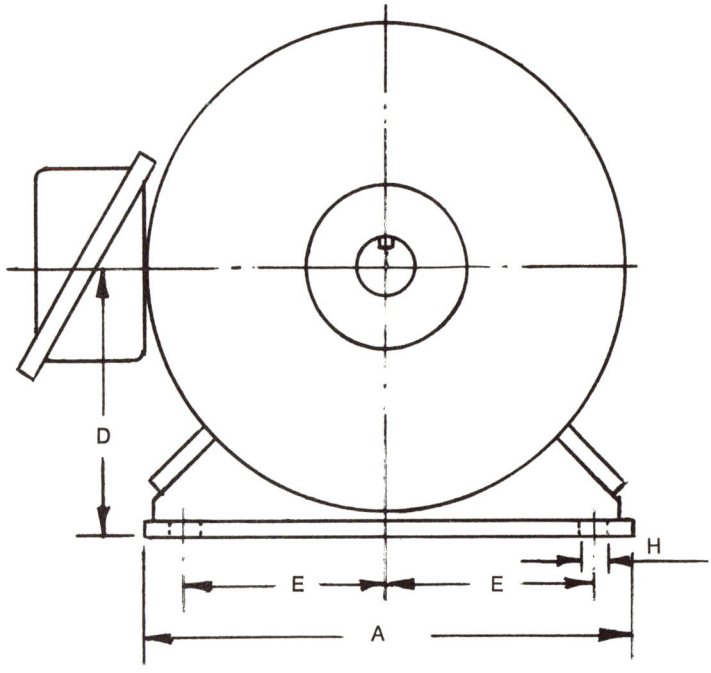

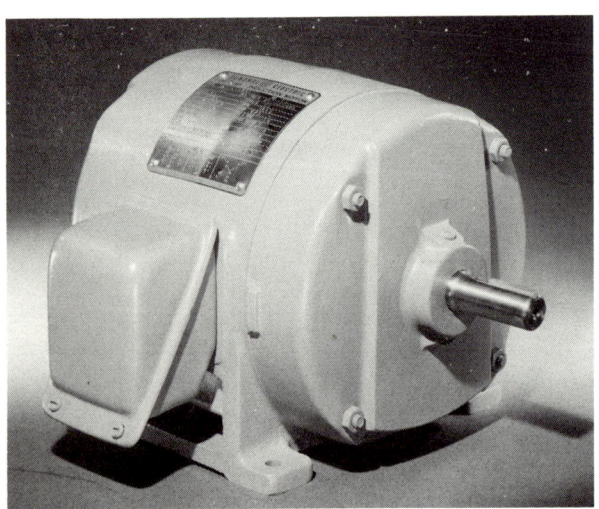

Figure 4.4–3. Foot mounted electric motor. (Courtesy of General Electric Co.)

4.4 Electrical Components

Table 4.4–2b. Dimensions for Foot Mounted a-c Motors *

Frame Number	A	D	E	H	Key width	Key thickness
182	9	4½	3¾	13/32	3/16	3/16
184	9	4½	3¾	13/32	3/16	3/16
213	10½	5¼	4¼	13/32	¼	¼
215	10½	5¼	4¼	13/32	¼	¼
254U	12½	6¼	5	17/32	5/16	5/16
256U	12½	6¼	5	17/32	5/16	5/16
284U	14	7	5½	17/32	3/8	3/8
286U	14	7	5½	17/32	3/8	3/8
324U	16	8	6¼	21/32	½	½
324S	16	8	6¼	21/32	3/8	3/8
326U	16	8	6¼	21/32	½	½
326S	16	8	6¼	21/32	3/8	3/8
364U	18	9	7	21/32	½	½
364S	18	9	7	21/32	½	½
365U	18	9	7	21/32	½	½
365S	18	9	7	21/32	½	½
404U	20	10	8	13/16	5/8	5/8

* Courtesy of General Electric Co.

very little with load and is close enough for most cases. The most common sizes for machine tools are from ½ to 15 hp.

2. Fractional horsepower a-c induction motors. These motors find uses in machine tools for auxiliary equipment, and functions of minor importance.
3. Torque motors. These motors have a high accelerating torque and the speed versus torque curve is approximately linear. They are therefore well suited for machine tools for motions of an

Table 4.4–2c. Dimensions for Face Mounted a-c Motors Type C *

Frame Number	AH	AJ	AK	BB	BD	BF Hole		
						Number	Tap Size	Bolt Penetration Allowance
182C	2⅛	5⅞	4½	5/32	6½	4	⅜-16	9/16
184C	2⅛	5⅞	4½	5/32	6½	4	⅜-16	9/16
213C	2¾	7¼	8½	¼	9	4	½-13	¾
215C	2¾	7¼	8½	¼	9	4	½-13	¾
254UC	3½	7¼	8½	¼	10	4	½-13	¾
256UC	3½	7¼	8½	¼	10	4	½-13	¾
284UC	4⅝	9	10½	¼	11¼	4	½-13	¾
286UC	4⅝	9	10½	¼	11¼	4	½-13	¾
324UC	5⅜	11	12½	¼	14	4	⅝-11	15/16
324SC	3	11	12½	¼	14	4	⅝-11	15/16
326UC	5⅜	11	12½	¼	14	4	⅝-11	15/16
326SC	3	11	12½	¼	14	4	⅝-11	15/16
364UC	6⅛	11	12½	¼	14	8	⅝-11	15/16
364USC	3½	11	12½	¼	14	8	⅝-11	15/16
365UC	6⅛	11	12½	¼	14	8	⅝-11	15/16

Figure 4.4–4. Face mounted electric motor. (Courtesy of General Electric Co.)

Frame Number	AH	AJ	AK	BB	BD	BF Hole		
						Number	Tap Size	Bolt Penetration Allowance
365USC	3½	11	12½	¼	14	8	⅝-11	15⁄16
404UC	6⅞	11	12½	¼	15½	8	⅝-11	15⁄16

° Courtesy of General Electric Co.

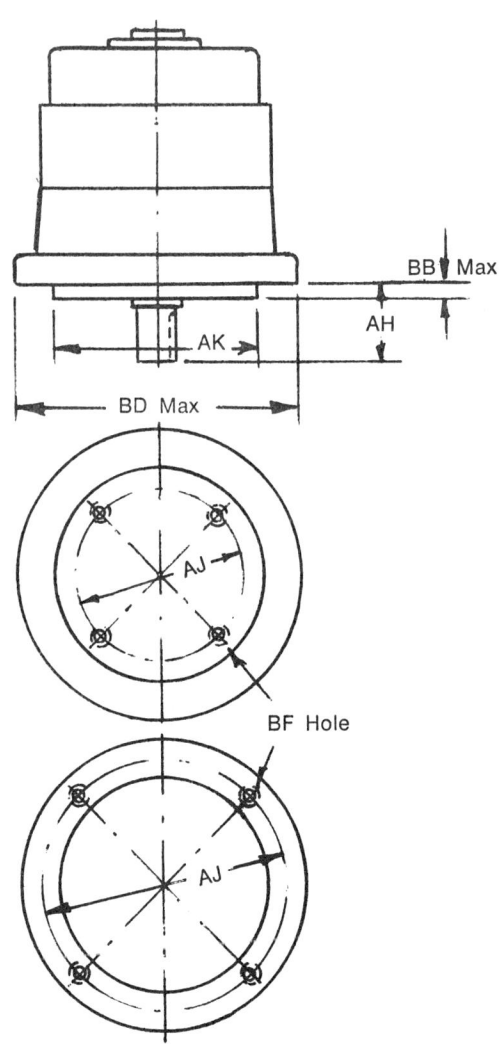

Prototype Machine

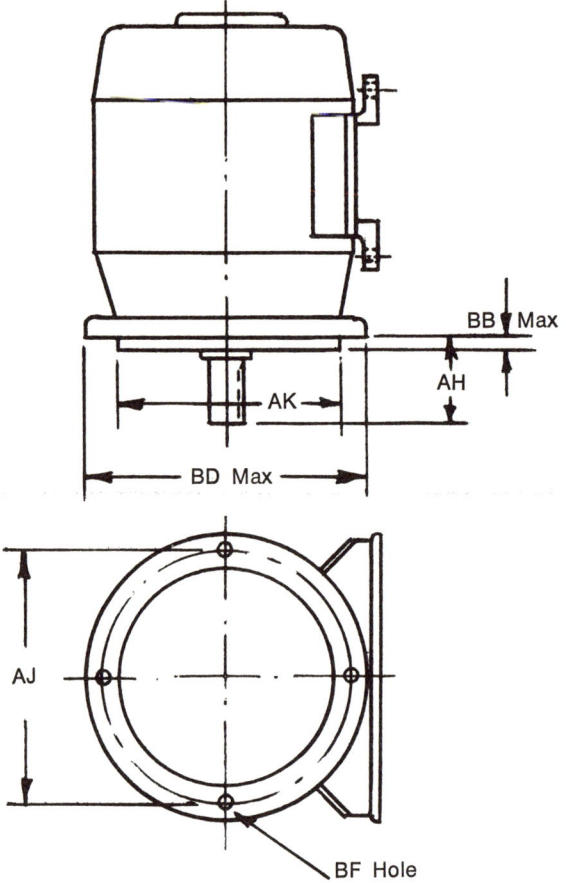

auxiliary nature, such as indexing mechanisms or clamping, or parts which have to be moved rapidly and then held firmly in place for some time.

4. Integral horsepower d-c motors. These motors have limited use for machine tools. They are, however, very flexible for speed variation for certain grinders or lathes when d-c is available.

5. Multispeed motors. These motors find some use in machine tools where it is desirable to change the speed in definite steps, as for instance, 3600 and 1800 rpm or 1800 and 900 rpm.

6. Slow speed a-c synchronous motors. These motors will run at

4.4 Electrical Components

Table 4.4–2d. Dimensions for Flange Mounted a-c Motors Type D *

Frame Number	AJ	AK	AH	BB	BD	BF Clearance Hole		Recommended Bolt Length
						Number	Size	
182D & 184D	10	9	2¼	¼	11	4	$17/_{32}$	1¼
213D & 215D	10	9	3	¼	11	4	$17/_{32}$	1¼
254UD & 256UD	12½	11	3¾	¼	14	4	$13/_{16}$	2
284UD & 286UD	12½	11	4⅞	¼	14	4	$13/_{16}$	2
324UD & 326UD	16	14	5⅝	¼	18	4	$13/_{16}$	2
324SD & 326SD	16	14	3¼	¼	18	4	$13/_{16}$	2
364UD & 365UD	16	14	6⅜	¼	18	4	$13/_{16}$	2
364USD & 365USD	16	14	3¾	¼	18	4	$13/_{16}$	2
404UD & 405UD	20	18	7⅛	¼	22	8	$13/_{16}$	2¼

* Courtesy of General Electric Co.

Figure 4.4–5. Flange mounted electric motor. (Courtesy of General Electric Co.)

Prototype Machine

Figure 4.4–6. Slow speed synchronous motor. (Courtesy of The Superior Electric Company)

72 rpm without gear reduction, will start and stop in approximately 1.5 cycles, which for 60 cps is 0.025 second. They will run at constant speed and can be instantly started, reversed and stopped with a three position single pole switch. There is no need for electrical or mechanical braking. (Figure 4.4–6)

Auxiliary Electrical Equipment

The control and operation of electrical circuits for machine tools is a highly specialized field, and is usually handled by the electrical engineering department, unless the design engineer is sufficiently familiar with electrical engineering to handle this himself.

Only some of the most common components will be mentioned here. Detailed information as to operating characteristics and physical dimensions of interest to the designer may be obtained from the manufacturers.

Switches

PUSH BUTTONS. These switches are pilot devices and are operated manually by the operator of the machine tool and should be located in a convenient location. (Figure 4.1–1) They should, however, be protected, so they cannot be operated accidentally. On some types of

4.4 Electrical Components

machine tools it may be advisable to have two push buttons located far enough apart so they must be operated by both hands simultaneously to start operation. This is to prevent the operator from getting his hands endangered by moving parts.

Most commonly, however, the push buttons are located on panels which also contain selector switches and lights. There are usually two main types of push buttons for machine tools:

1. Momentary contact type used for controlling magnetic con-

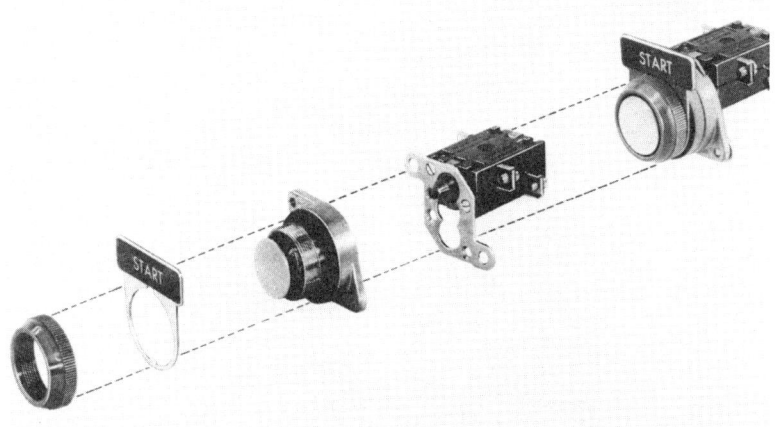

Figure 4.4–7a. Push button unit adapts itself very well to a panel mounting. (Courtesy of General Electric Co.)

Figure 4.4–7b. Push button suitable for designs where two units are used far enough apart so that the two buttons must be operated simultaneously to start the machine. (Courtesy of General Electric Co.)

Prototype Machine

Figure 4.4–7c. Push button suitable for light machine tools. (Courtesy of General Electric Co.)

trollers located in the main electrical panel compartment. The contacts for these switches are closed or opened only while the button is manually operated.
2. Maintained contact type used as a selector switch to select a certain control circuit. The contacts for these switches remain closed or open after the button has been manually operated.

SELECTOR SWITCHES. The most common selector switches for machine tools are of the rotary type. They are usually operated to one of two positions, run or jog, on or off, manual or automatic.

Lights of different colors are also usually placed close to the switches on the push button panel to indicate to the operator normal running operation, cycle changes, or a warning. (Figures 4.4–7a, 4.4–7b and 4.4–7c)

LIMIT SWITCHES. These switches are most commonly operated by mechanical means from levers, slides, clamping devices, cams or other motions in the machine to send an electrical impulse to a relay in the electrical control compartment which will close, alter or memorize a circuit for later use. The machine may thus be designed with limit switches in proper locations to maintain an orderly sequence of opera-

4.4 Electrical Components

tions, making sure that each operation is satisfactorily finished before the next one is started. These switches should be of the snap action type so their operation is not dependent on the speed of the operating means. They are usually designed for a certain minimum pretravel, minimum differential travel or operating travel, and minimum overtravel. For the push type, the maximum overtravel is important.

The differential or operating travel may be as small as 0.0005 inch. The force required to operate the switch may range from 2 to 30 ounces, depending on type of switch. There are two main kinds of mechanically operated switches for machine tools: the lever and roller limit switch (Figures 4.4–8 and 4.4–9), and the push limit switch (Figure 4.4–10).

Other types of switches are operated by flow or pressure of liquids or gases, or by temperature. Another, the mercury switch, is well adapted to angular motions in a vertical plane and is insensitive to ambient temperature conditions.

Linear Electrical Actuators

There are two main types—solenoids and polenoids. Solenoids, which are the most widely used, push or pull a machine member a

Figure 4.4–8. Limit switch, mechanically operated by roller and lever arrangement. (Courtesy of R. B. Denison, Inc.)

Prototype Machine

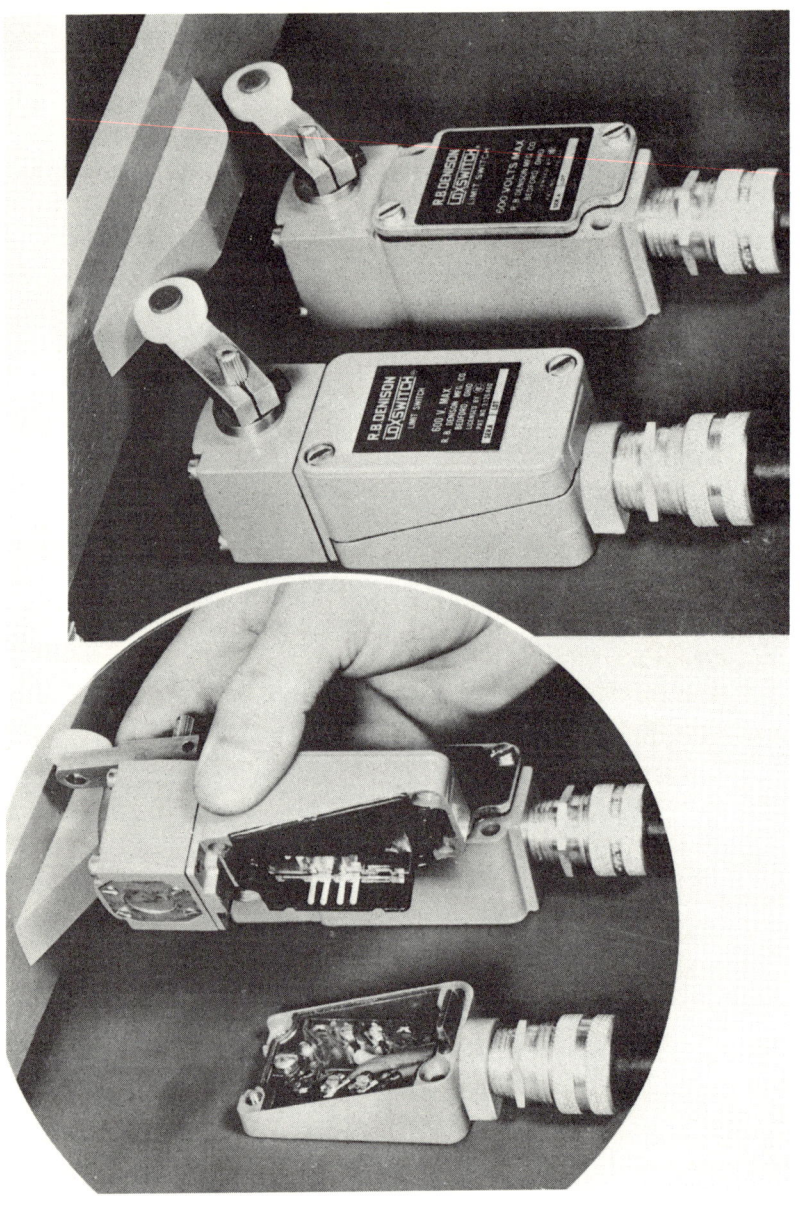

4.4 Electrical Components

Figure 4.4–10. Mechanically operated push type limit switch. (Courtesy of R. B. Denison, Inc.)

comparatively short distance. The application should be designed in such a way that the solenoid is completely seated to protect the coil from overheating and burning out and to cut down the noise. (Figure 4.4–11) Polenoids, which are fundamentally linear induction motors, provide a long stroke of a uniform force. Length of stroke is electrically unlimited, and reaction to energization is less than 10 milliseconds. Normal operating speed for a three phase unit is approximately 90 ips. There are basically two parts to the unit; the stator containing the windings, and the moving part, the rod. (Figure 4.4–12) Polenoids are useful as long linear actuators for light loads when air or hydraulics are not available.

Figure 4.4–9. Limit switches actuated by cam on machine tool in contact with roller and lever arrangement on limit switch. If the limit switch is subject to high speeds and frequency of operation, where it would be necessary to replace the switch without loss in production, the designer should consider a plug in switch. Such a switch is shown on the left side of this illustration. The electrical connection of the switch to the wiring system is broken in the same manner as when an ordinary plug is removed from a receptacle, thus no wiring is necessary. (Courtesy of R. B. Denison, Inc.)

Prototype Machine

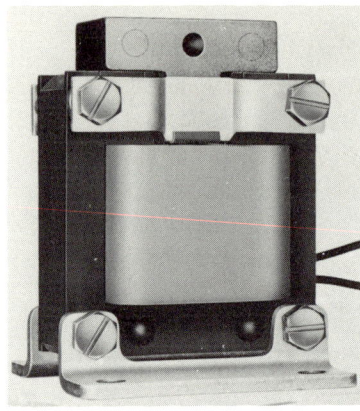

Figure 4.4–11. Solenoids are used for comparatively short strokes, and because the work obtained from a solenoid is constant, it should be selected for a stroke length and load close to the requirements. Otherwise the unnecessary length of stroke and unnecessary energy above requirements not used in overcoming the external resistance will result in a heavy impact on the solenoid frame and also will be dissipated in the form of heat. (Courtesy of National Acme Co.)

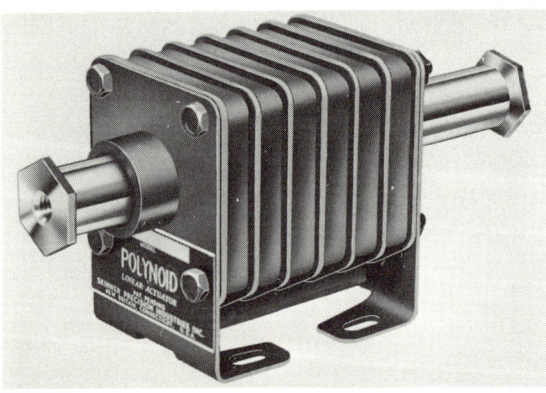

Figure 4.4–12. Polenoid, or an electrical linear actuator. (Courtesy of Skinner Precision Industries, Inc.)

Transducers and Potentiometers

These are instruments for measuring electrically amounts of mechanical deflection, or elongation, or temperature changes in a pyrometer. Very accurate measurements of deflection or mechanical changes may be made electrically from a strain gage secured to the part or parts being measured, and the measurements may be performed when the parts are under motion.

4.5 Pneumatics

Relays, Timers and Counters

RELAYS. Most of these components are located in the electrical panel compartment. One type provides a time delay before or after operation, and is therefore a type of timer. Other types keep the machine operations in the proper sequence through push buttons or limit switches.

TIMERS. There are many types of timers:

1. Mechanically operated
2. Pneumatically or hydraulically operated
3. Thermally operated
4. Motor driven
5. Electrically operated

Counters

These instruments are very useful for machine tools. They are available in several different types and can be used for counting to give an accurate record of production; or to stop a machine at the end of a preset number of impulses.

4.5 Pneumatics in Machine Tool Design

Compressed air is used in almost all manufacturing plants. The power derived from compressed air is versatile, and finds many useful applications to machine tools. The engineering student or practicing engineer, may, however, be so engrossed in the mechanical aspects or functions of the machine as a whole that he gives little attention to the proper selection and location of the pneumatic components.

A thorough understanding of the fundamental characteristics of air at different temperatures and pressures is necessary for successful application of pneumatics to machine tools.

Dry air at sea level is composed of 78.03% nitrogen, 20.99% oxygen and 0.98% argon. All air carries some water vapor. The amount of water vapor carried in a certain volume of air depends largely on the temperature of the air. The amount of water vapor in free air is only slightly influenced by pressure. It may, however, be of interest to notice, that it takes 9.65 cu ft of air at sea level, to make 1 cu ft of compressed air at 100 psi. The amount of water vapor that is carried by a volume of air is given as relative humidity. At high temperature, air will carry a larger amount of water vapor than at lower tempera-

Prototype Machine

ture. The amount of water vapor the air will carry at 100% relative humidity approximately doubles for every 20°f temperature rise. When the relative humidity reaches a point of saturation (100%), the water vapor is released as a liquid.

It is, therefore, easily understood that if an air line is brought down to a machine tool from ceiling level, there may be several degrees difference in temperature, and as a result, the air is cooled off rapidly, can no longer retain the water vapor, and releases it as liquid. Only the water above the point of saturation is released. To release any water vapor below the point of saturation, chemical or other drying means must be employed. For all practical purposes, air below 100% relative humidity is considered dry air, and is satisfactory for most machine tool applications.

Water is injurious to pneumatic components used on machine tools, so proper provisions should be made to get rid of the water that is released before it reaches the pneumatic components.

First, let us consider the requirements for air supply to the machine tool. As a rule, many machines use pneumatic components in a machine shop. It is, therefore, desirable to have one compressor take care of a group of several machines, if the demand is heavy. (Figure 4.5–1) The air supply and pressure must be adequate to meet the requirements of all machines. It is advisable to have a supply tank of sufficient capacity to meet the heaviest demand expected. This tank is also called a receiver tank.

Free air is the pressure of the air at the intake of the compressor and varies with the altitude. It is 14.7 psia at 62°f, or in the metric system 1 atmosphere at sea level. In ordinary machine tool engineering work, the pressure is measured above the atmospheric pressure as gage pressure (psig). The gage is set at zero reading at atmospheric pressure and any pressure read on the gage is known as psi to the design engineer when calculating forces. The absolute pressure equals the atmospheric pressure plus the gage pressure, 14.7 + psig = psia.

For application of pneumatic components to machine tools, it is not necessary to be familiar with the more complicated laws of adiabatic or isothermal expansion and compression of air. One thing

Figure 4.5–1. Typical pneumatic installation layout. This is a schematic layout. In planning application to machine tools, consideration should be given to easily accessible components, well protected from abuse from other shop equipment.

4.5 Pneumatics

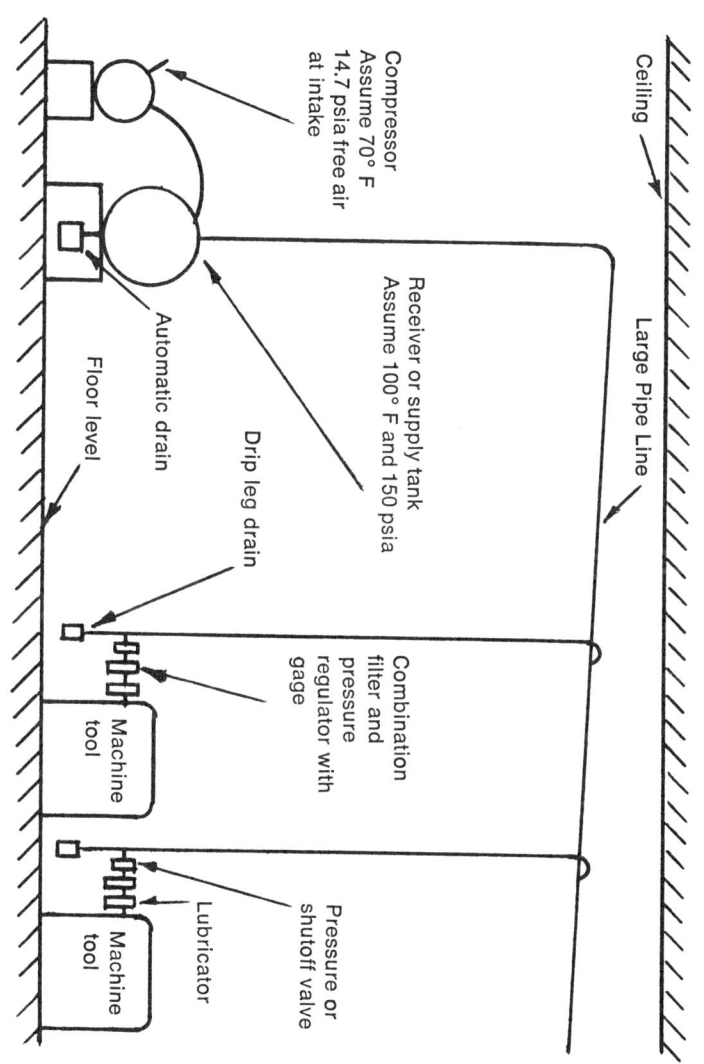

Prototype Machine

should, however, be noticed that under constant pressure, the volume is directly proportional to absolute temperature. If, on the other hand, the temperature remains constant, the volume is inversely proportional to the absolute pressure.

From the gas laws we have this formula:

$$\frac{P \times V}{T} = M \times R$$

Assume then that:

$M = 1$ lb, weight of air mixture
$R = 53.35$, gas constant for air
$P =$ absolute partial pressure of air in pounds per square foot
$T =$ absolute temperature in degrees fahrenheit
V_a or $V_v =$ volume of air in cubic feet; this is also the volume of the vapor in the air, since vapor occupies the same space.

Referring to figure 4.5-1, we will then assume it is required to calculate the amount of water in gallons that may be released at the machine level in 8 hours by the air system.

$P = 14.7$ psia, free air, or atmospheric pressure at intake of compressor
$P_{1t} = 150$ psia, absolute pressure of air at 100 degrees fahrenheit
$P_{2t} = 0.9492$ psia, absolute partial pressure of the vapor at 100 degrees fahrenheit, from steam tables
$P_t = p_{1t} - p_{2t} = (150 - 0.9492) \times 144 = 21{,}463$ psf
$t_t = 100°$f, temperature of air mixture at tank
$T_t = 459.6 + 100 = 559.6°$f
$s_t = 350.3$, saturated vapor at 100°f from steam tables
$d_t = \dfrac{1}{350.3} = 0.00285$ pounds per cubic foot
$V_{at} = \dfrac{M \times R \times T_t}{P_t} = \dfrac{1 \times 53.35 \times 559.6}{21{,}463} = 1.390$ cu ft,
volume of air at tank for 1 lb of air mixture
$w_t = d_t \times V_{at} = 0.00285 \times 1.390 = 0.00396$ lb, weight of saturated vapor in the air mixture
$V_{vt} = 1.390$ cu ft, volume of vapor at tank for 1 pound of air mixture

4.5 Pneumatics

Since the air and vapor occupy the same space, we have:
$$\frac{w_t}{V_{vt}} = \frac{0.00396}{1.390} = 0.00285 \text{ pounds of water vapor per cubic foot of air}$$
mixture. As one pound is 7000 grains, this then is: $0.00285 \times 7000 = 19.95$ grains of water vapor at saturation, carried by system at the tank for 1 cubic foot of air mixture at 100°f and 150 psia.

Now, if we follow the same procedure for saturated vapor or 100% relative humidity at the machine intake level, we have:

$p_g = 90$ psig, gage pressure at the machine
$p_{1m} = p_g + p = 90 + 14.7 = 104.7$ psia
$p_{2m} = 0.3631$ psia, absolute partial pressure of the vapor at 70°f, from steam tables
$p_m = (104.7 - 0.3631) \times 144 = 15,025$ psf
$t = 459.6$°f, absolute zero temperature
$t_m = 70$°f, temperature of air mixture at machine
$T_m = 459.6 + 70 = 529.6$°f
$s_m = 867.8$, saturated vapor at 70°f, from steam tables
$d_m = \dfrac{1}{867.8} = 0.001152$ pounds per cubic foot
$V_{am} = \dfrac{M \times R \times T_m}{P_m} = \dfrac{1 \times 53.35 \times 529.6}{15,025} = 1.88$ cu ft, volume of air at the machine for one pound of air mixture
$w_m = d_m \times V_{am} = 0.001152 \times 1.88 = 0.00217$ lb, weight of saturated vapor in the air mixture
$V_{vm} = 1.88$ cu ft, volume of vapor at the machine for 1 pound of air mixture

Since the air and vapor occupy the same space, we have:

$$\frac{w_m}{V_{vm}} = \frac{0.00217}{1.88} = 0.001152 \text{ pound per cubic foot}$$

Since one pound is 7000 grains, this then is: $0.001152 \times 7000 = 8.064$ grains of vapor at saturation carried by the system at the machine level for 1 cubic foot of air mixture at 70°f and 90 psig.

Since 19.95 grains of vapor per cubic foot of air mixture is supplied to the machine from the tank, we have: $19.95 - 8.064 = 11.886$ grains of vapor which will drop out as water at machine intake level.

Prototype Machine

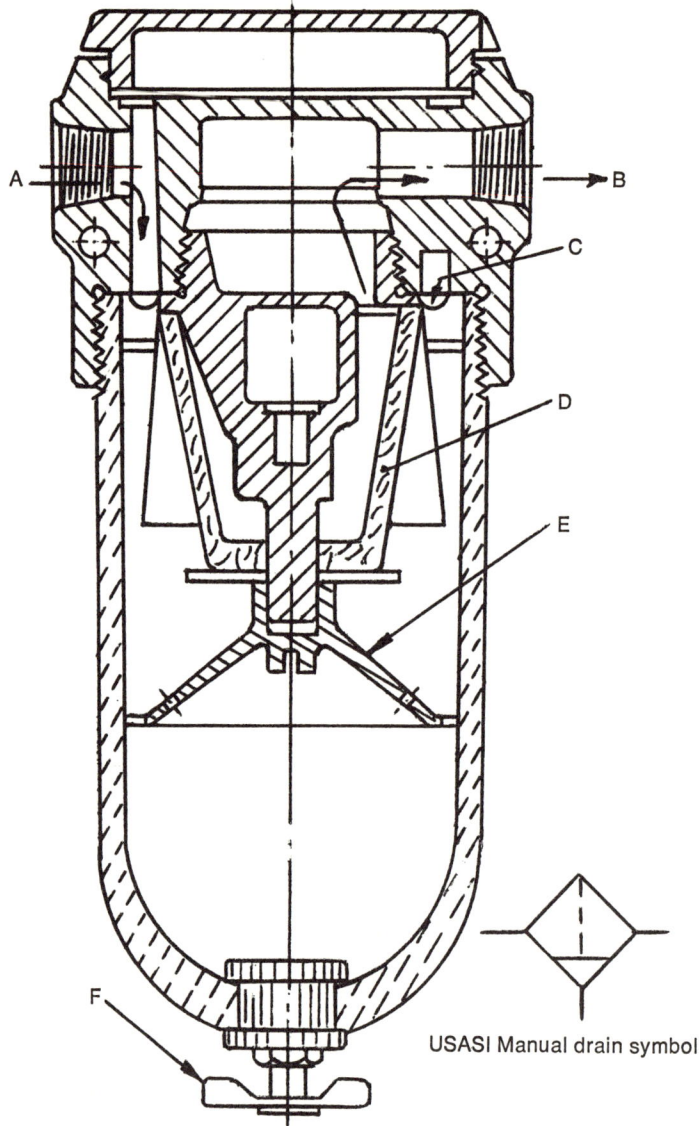

USASI Manual drain symbol

Figure 4.5–2a. Air filter operation. Air flows in from port A through the directional louvers C, forcing it into a whirling flow pattern. Liquid particles are thrown against the inside wall of the bowl by centrifugal force. The liquid particles run down into the bottom of the bowl. The baffle E maintains a quiet zone in the lower part of the bowl to prevent air turbulence from picking up the liquid and returning it to the air stream. The air then passes through the filter element D to remove solid contaminants. Liquid contaminants are drained by opening the manual drain cock F.

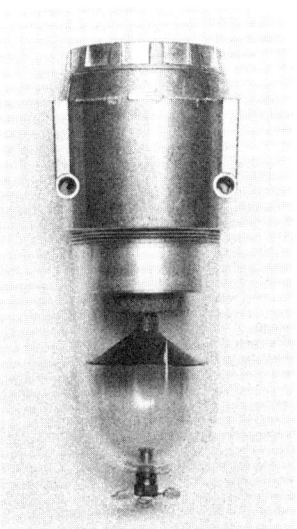

Figure 4.5-2. Air filter. (Courtesy of C. A. Norgren Company)

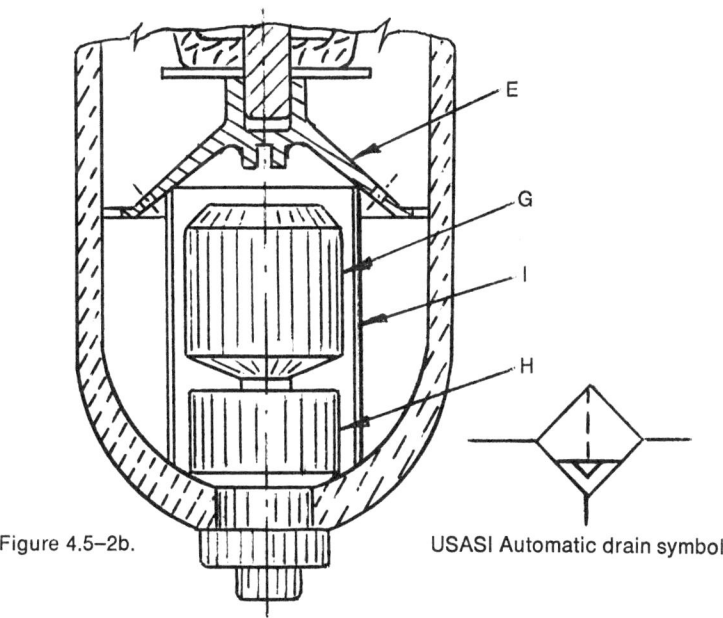

Figure 4.5-2b. USASI Automatic drain symbol

An automatic drain assembly, Figure 4.5-2b, easily interchangeable with the manual drain, automatically dumps liquid as it collects. As liquid builds up in the bowl, float G rises, causing piston type automatic drain assembly H to open and release the liquid under pressure. Liquid may also be dumped manually by depressing needle inside drain outlet. Screen I protects drain mechanism.

Standard filters are usually made to filter out particles of 50 microns = .002 inch. (Courtesy of C. A. Norgren Company)

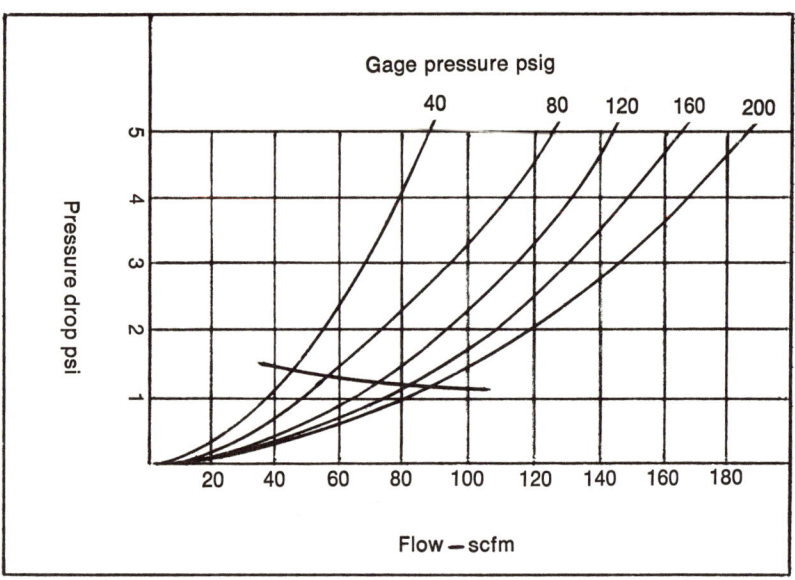

Figure 4.5–2c. Filter flow characteristics
Maximum recommended flow for ½ inch standard pipe size and 50 micron filter element is indicated where psig values intersect curve. (Courtesy of C. A. Norgren Company)

In an 8 hour day, if the air consumption at the machine is 100 cfm, we would have:

$$\frac{11.886 \times 60 \times 8 \times 7.481}{7000 \times 64} = .9527 \text{ gallon}$$

Therefore, to prevent moisture from doing damage to the machine components, provisions should be made for removing that amount of water per day.

It is also recommended that the air be filtered for removal of compressor oil and foreign objects such as scales that may have accumulated at the compressor, and that the accumulated water be drained from the system before the air enters the pneumatic components. (Figures 4.5–2, 4.5–3, 4.5–4a) There should be no substantial drop in the supply line after takeoff from the filter to the pneumatic components. (Figure 4.5–1) Because all moving parts need to be lubricated, a lubricator should be installed in the line to inject atomized lubricant into the air stream. (Figure 4.5–3 and 4.5–5)

4.5 Pneumatics

Figure 4.5–3. Combination filter and pressure regulator with lubricator.

The combination filter and pressure regulator filters liquid and solid contaminants from the air. It also maintains a constant pressure to the pneumatic components as indicated by the gage.

The lubricator operation is shown in figure 4.5–5. (Courtesy of C. A. Norgren Company)

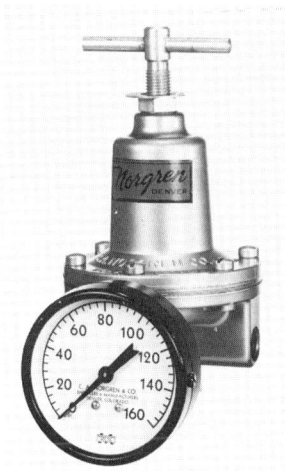

Figure 4.5–4a. Air pressure regulator. (Courtesy of C. A. Norgren Company)

133

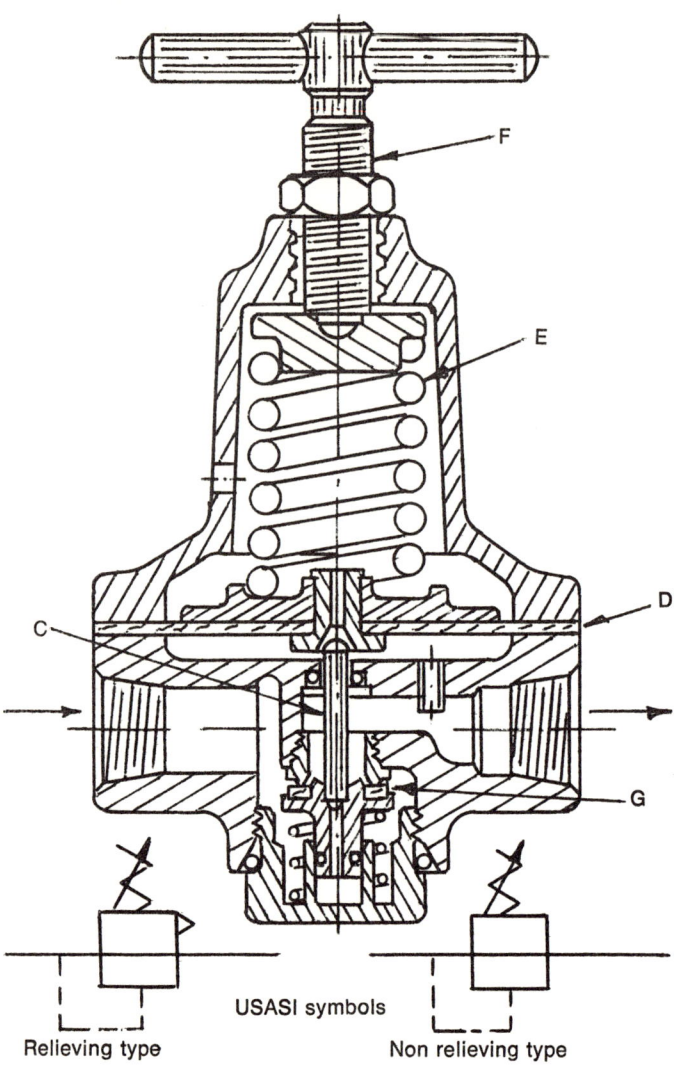

4.5 Pneumatics

Figure 4.5–4b. Pressure regulator operation. The working elements of a pressure regulator consist mainly of a flexible diaphragm D, which controls a valve through an interconnecting pin C, and an adjustable spring E which is loaded by means of an adjusting screw F.

The primary pressure enters the regulator through port A. The pressure side of the diaphragm is connected to the outlet port B of the regulator, so that the secondary or regulated pressure will be exerted against the diaphragm.

When the adjusting screw is retracted, so that no load is applied to the adjustable spring, the regulator valve G is closed. As the adjusting screw is turned in, it applies a load to the adjustable spring, which is transmitted to the valve through the diaphragm and the valve pin, thus opening the valve. As the regulated pressure increases, the pressure against the diaphragm increases, forcing the diaphragm to compress the adjustable spring until the load exerted by the adjustable spring is equal to the load exerted by the regulated pressure. If there is no flow demand, this state of equilibrium will occur with the valve closed. If there is a flow demand, this state of equilibrium will occur with the valve open just the amount necessary to compensate for the demand, thus maintaining the desired regulated pressure as indicated on the gage. (Courtesy of C. A. Norgren Company)

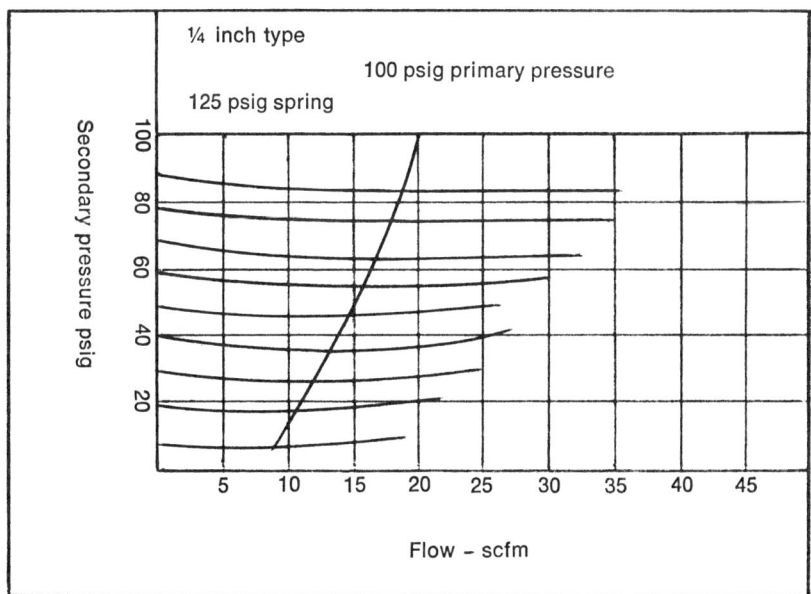

Figure 4.5–4c, 4d and 4e. Pressure regulator flow characteristics. Maximum recommended flow, for size of pipe system used, may be found, where secondary pressure curve used, intersects oblique curve. (Courtesy of C. A. Norgren Company)

135

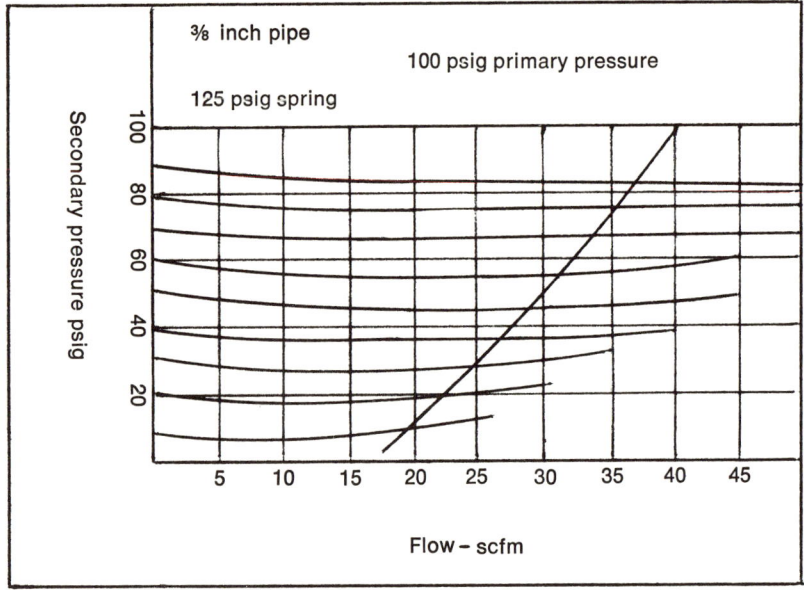

Figure 4.5-4d

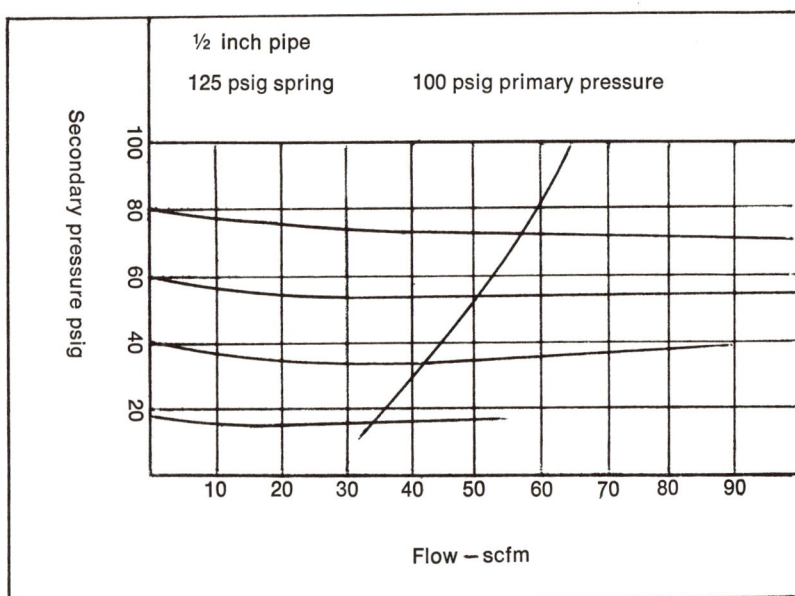

Figure 4.5-4e.

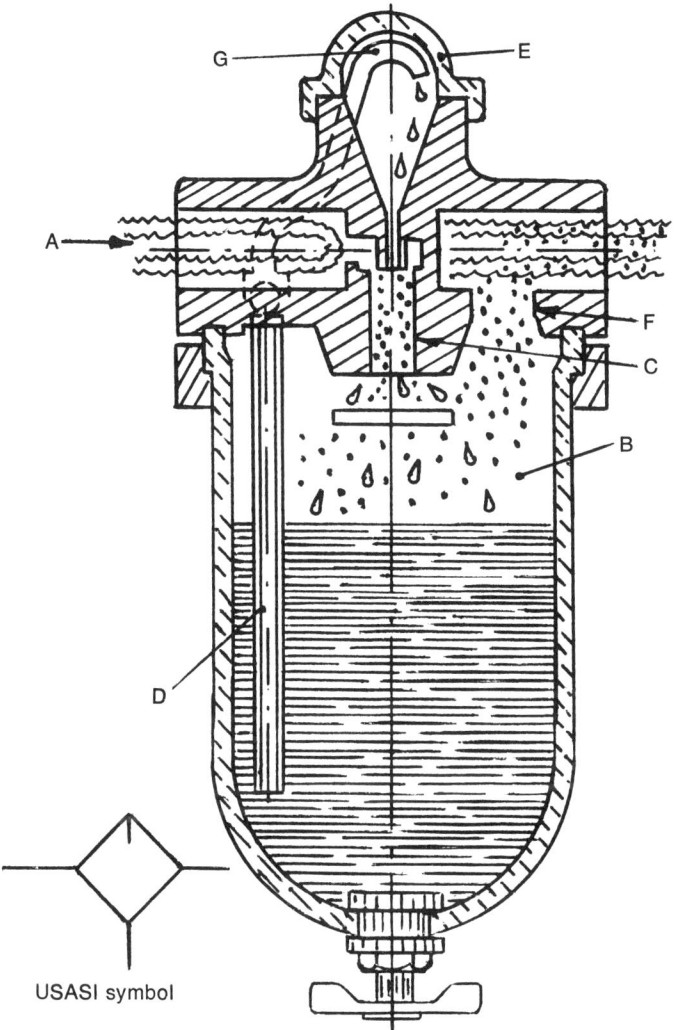

Figure 4.5–5. Micro fog lubricator operation
Part of the air entering inlet A is directed through a venturi into the oil reservoir B. The remaining air is diverted around a vane bypass, which is adjustable for the desired micro fog output to suit the air flow. The air passing through the venturi C creates a pressure differential which causes the oil in the reservoir to flow up through the siphon tube D to the sight feed dome E and drip into the venturi C. For every 20 drops passing through the drip tube G, one drop enters the air line as micro fog, making possible extremely accurate adjustment. The action of the air and oil at the venturi creates a finely divided oil fog in the upper part of the oil reservoir B. All oil particles larger than 2 microns fall out, returning to the oil reservoir. The smaller particles still remain airborne, and represent about 5% of the oil which passed through the sight feed dome. The micro fog then travels through passageway F to the air line and multiple points of lubrication. (Courtesy of C. A. Norgren Company)

Prototype Machine

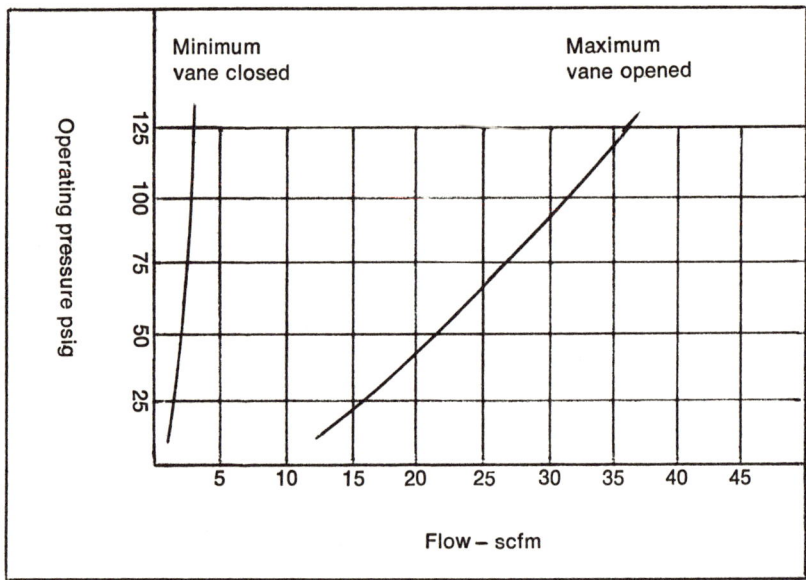

Figure 4.5–5a. Micro fog lubricator performance characteristics. The lubricator operating range lies between the two oblique curves. This is based on ¼ inch and ⅜ inch pipe sizes using SAE 10 oil. Maximum oil flow setting is based on 5 psi pressure drop. (Courtesy of C. A. Norgren Company)

Make all the air lines as large in diameter as possible for there is a considerable drop in pressure in smaller-diameter lines. (Table 4.5–6)

Maximum air pressure in most plants is 90 psi, and most tools are designed for 80 to 90 psi, so every precaution should be taken to make sure that the pressure at the machine will be within this range to keep the efficiency of the tools at a high level. Also make sure that all valves are of sufficient size to carry the volume required. (Table 4.5–7 and 4.5–8)

Some components are available for sub base mounting. (Figures 4.5–9, 4.5–10a and 10b) This is a good feature because the parts may be removed for replacement or repair without removing pipes or wires. If you have several components close together, you can design a manifold, a large plate drilled or channeled to suit the components. (Figure 4.5–11) Only one pressure line inlet is then necessary, and if mufflers are required, one may be sufficient. Make sure that all drilled holes are as large as possible. If a channeled manifold is used, it may be made in two parts and the channels may be cast with large radii. Because the

Table 4.5–6. Loss of Air Pressure Due to Friction *

Per 100 Ft. of Pipe and 100 psi Initial Pressure

cfm Free Air	Equivalent cfm Comp. Air	Nominal Pipe Diameter in Inches							
		½	¾	1	1¼	1½	2	2½	3
10	1.28	.38	.09	.03	.007				
20	2.56	1.42	.34	.10	.026	.012			
30	3.84	3.13	.74	.23	.056	.026			
40	5.13	5.55	1.28	.38	.096	.044	.013		
50	6.41	8.65	2.00	.60	.146	.067	.020	.008	
60	7.69		2.84	.84	.21	.095	.027	.011	
70	8.97		3.85	1.12	.28	.130	.036	.015	
80	10.25		5.01	1.44	.36	.160	.046	.019	
90	11.53		6.40	1.85	.45	.200	.058	.024	
100	12.82		7.80	2.21	.55	.250	.069	.029	.010
125	16.02		12.40	3.41	.85	.380	.107	.043	.015
150	19.22		18.10	4.91	1.20	.54	.150	.061	.021
175	22.43			6.80	1.64	.73	.200	.081	.028
200	25.63			8.79	2.12	.95	.260	.105	.036
250	32.04				3.30	1.48	.400	.160	.054
300	38.45				4.71	2.10	.570	.230	.075
350	44.86				6.45	2.86	.77	.310	.101
400	51.26				8.30	3.70	.99	.400	.131
450	57.67					4.65	1.27	.500	.165
500	64.08					5.79	1.56	.620	.200
600	76.90					8.45	2.23	.890	.290
700	89.71						3.00	1.180	.390
800	102.50						4.00	1.540	.500
900	115.30						5.05	1.950	.630
1,000	128.20						6.20	2.370	.780

* From Wilkerson Corporation.

Table 4.5-7. Flow of Air Through Orifice in CFM *

Supply Pressure psi	Orifice Size									
	1/32	1/16	3/32	1/8	5/32	3/16	7/32	1/4	9/32	5/16
2	.17	.62	1.44	2.55	3.93	5.74	7.71	10.20	13.00	15.90
5	.25	.94	2.19	3.77	5.95	8.70	11.60	15.40	19.60	24.00
10	.39	1.48	3.41	6.10	9.31	13.60	18.20	24.10	30.60	37.50
15	.41	1.62	3.63	6.44	10.00	15.40	19.60	25.80	32.60	40.00
20	.49	1.90	4.28	7.59	11.80	17.00	23.10	30.40	38.40	47.20
25	.56	2.18	4.93	8.74	13.60	19.60	26.60	35.00	44.20	54.30
30	.63	2.47	5.58	9.89	15.40	22.20	30.40	39.60	50.10	61.50
35	.71	2.76	6.23	11.00	17.20	24.80	34.00	44.20	55.90	68.60
40	.78	3.05	6.88	12.20	19.00	27.30	37.50	48.80	61.70	75.80
45	.85	3.34	7.53	13.30	20.80	29.90	41.00	53.40	66.60	82.90
50	.92	3.62	8.17	14.50	22.60	32.50	44.60	57.00	72.40	90.10
55	1.00	3.91	8.82	15.60	24.40	35.10	48.10	62.70	78.20	97.30
60	1.07	4.20	9.47	16.80	26.20	37.70	51.60	67.20	84.10	104.40
65	1.15	4.49	10.10	17.90	27.90	40.30	55.20	71.80	89.90	111.70
70	1.21	4.77	10.80	19.10	29.70	42.80	58.80	76.40	95.70	118.80
75	1.30	5.06	11.40	20.20	31.50	45.40	62.30	81.00	105.50	126.00
80	1.37	5.35	12.10	21.10	33.30	48.00	65.80	85.60	107.40	133.10
85	1.44	5.64	12.70	22.50	35.10	50.60	69.40	90.30	113.20	140.30
90	1.52	5.92	13.40	23.70	36.90	53.20	72.90	94.80	119.00	147.50
95	1.59	6.21	14.00	24.80	38.70	55.70	76.50	99.40	124.90	154.60
100	1.66	6.50	14.70	26.00	40.50	58.30	80.00	104.60	130.70	161.80
125	2.03	7.94	17.90	31.70	49.50	71.40	97.70	127.10	159.80	197.50
150	2.40	9.28	21.20	37.50	58.40	84.40	115.40	150.10	189.00	233.30

* From Wilkerson Corporation.

4.5 Pneumatics

Table 4.5–8. Friction Loss in Pipe Fittings *

Figures Are Equivalent Ft. Straight Pipe, See Table 4.5–6

Type of Fitting	Nominal Pipe Size in Inches							
	¼	⅜	½	¾	1	1¼	1½	2
Gate valve, full open	.30	.30	.35	.44	.56	.74	.86	1.10
Tee, straight through	.50	.50	.70	1.10	1.50	1.80	2.20	3.00
Tee, side outlet	2.50	2.50	3.30	4.20	5.30	7.00	8.10	10.40
90° Ell	1.40	1.40	1.70	2.10	2.60	3.50	4.10	5.20
45° Ell	.50	.50	.78	.97	1.23	1.60	1.90	2.40
Angle valve, full open	8.00	8.00	9.30	11.50	14.70	19.30	22.60	29.00
Gate valve, full open	14.00	14.00	18.60	23.10	29.40	38.60	45.20	58.00

* From C. A. Norgren Company.

Figure 4.5–9. Sub-base mounted 4-way valve. (Courtesy of C. A. Norgren Company)

Prototype Machine

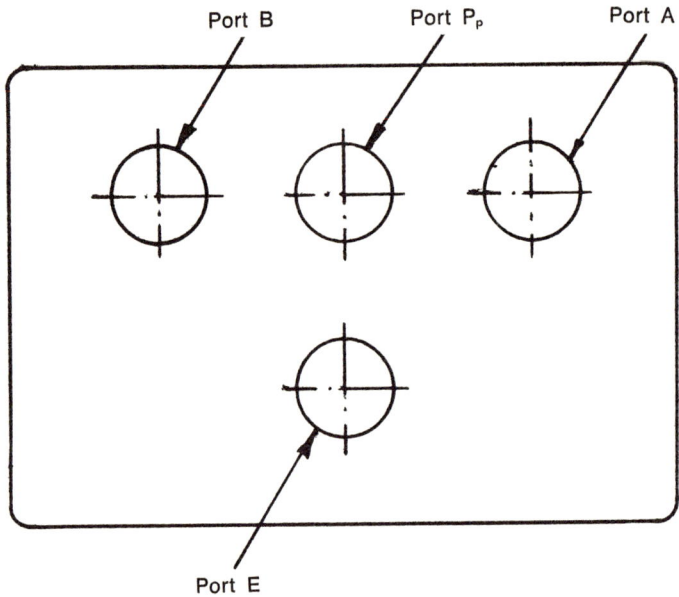

Figure 4.5–10a. 4-way valve

The 4-way valve is the most commonly used for operating actuators and cylinders in machine tools. For simplicity of illustration, the port holes are shown as above.

Maintained signal valve, single solenoid or single signal is used for operating double acting cylinders or similar components, especially where rapid action is required. It may be operated manually, mechanically, pneumatically or electrically by a pilot valve. It will hold the cylinder or actuator in the acquired position as long as the applied force to the valve is maintained. This is called a maintained signal. As soon as the applied force is released, the valve may be returned to its original position by a spring.

Momentary signal valve, operated by two solenoids, is useful when it is required to operate the system by the electrical pulse from a limit switch. The valve may then be kept in either operated position for a long duration of time without the requirement for continuous electrical energy. This valve is also used for operating double acting cylinders. One signal will put one end of the cylinder under pressure and the other end on exhaust. Another signal will reverse the system.

This valve may also be used to operate two single acting cylinders, clamping devices or ratchet operations. One signal would then put one cylinder under air pressure and let the other cylinder return by spring pressure. Another signal would put the other cylinder under air pressure and let the first cylinder return by spring pressure.

4.5 Pneumatics

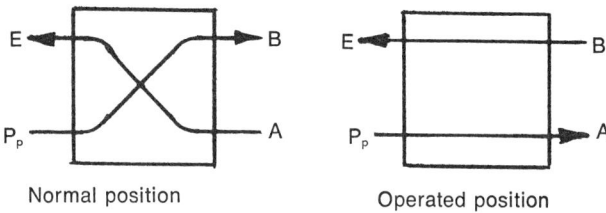

Factual observation

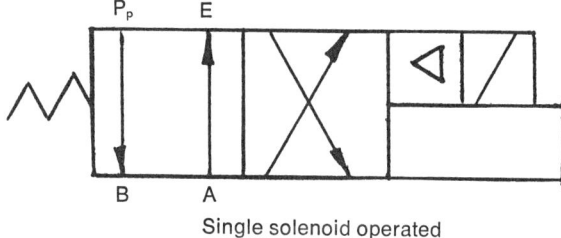

Single solenoid operated

USASI symbols

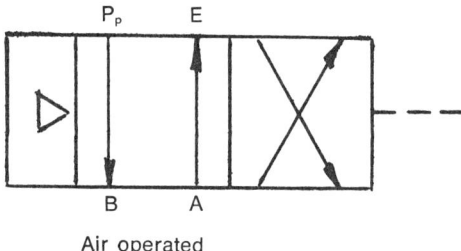

Air operated

Figure 4.5–10b. Operation of 4-way valve, maintained signal, single solenoid or single signal. See Figure 4.5–10a.

When pressure is applied to port P_p from the pilot valve, normal position, port A is exhausted through port E. Port B is pressurized from port P_p. In the operated position, port A is pressurized from port P_p and port B is exhausted through port E.

Prototype Machine

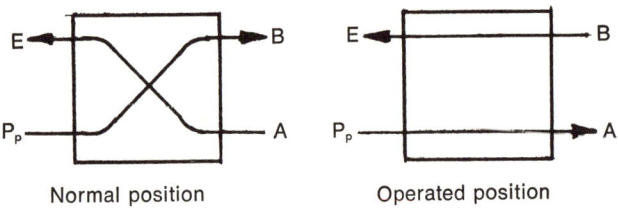

Normal position   Operated position

Factual observation

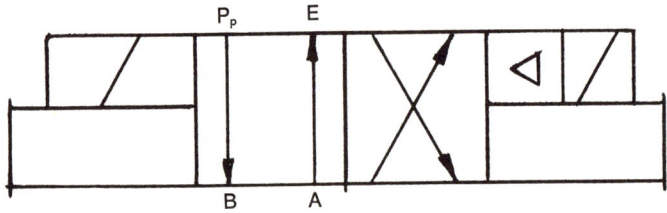

USASI symbol double solenoid operated

Figure 4.5–10c. Operation of 4-way valve, momentary signal, with two solenoids.
When pressure is applied to port P_p, after a momentary signal is applied to solenoid B, port A is exhausted through port E. Port B is pressurized from port P_p and remains pressurized until solenoid A receives a momentary signal. This then puts port A under pressure from port P_p. Port B is exhausted through port E.

two parts are permanently fastened together, the surfaces may be sealed with a gasket or plastic compound. The holes in the top of the plate, to mate the pneumatic components, may have counterbored holes for o-ring seals. (Figure 4.5–12)

A compressed air system properly planned, filtered and regulated finds several uses in machine tools. It may be an advantage to use air as power. Air gives you smooth acceleration and a cushioned stop is readily available. Such power is all right if the pneumatic components can be working while other mechanical work is being performed. If, on the other hand, the time cycle is dependent on response from the pneumatic components, make sure that supply and pressure are sufficient to meet the requirements. You, as a designer, are responsible for this decision.

Only some of the most common applications of pneumatic power to machine tools will be mentioned here. Many standard parts are

4.5 Pneumatics

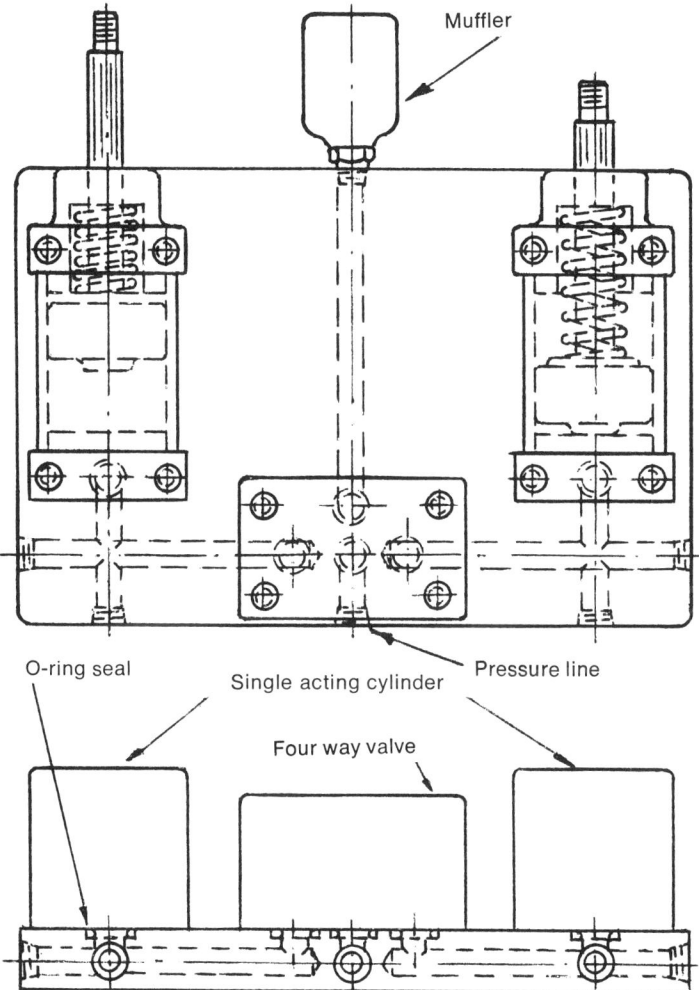

Figure 4.5–11. Manifold plate, drilled construction

One pilot operated 4-way valve, described in Figures 4.5–10a to 4.5–10c, operates two single acting cylinders. Many more components may be fastened to the same plate. It can be arranged so that only one pressure line is required, and if noise from the exhaust is objectionable, one muffler can serve the whole system.

If any of the components must be removed for replacement or service, this can be done without disturbing the pressure line.

Prototype Machine

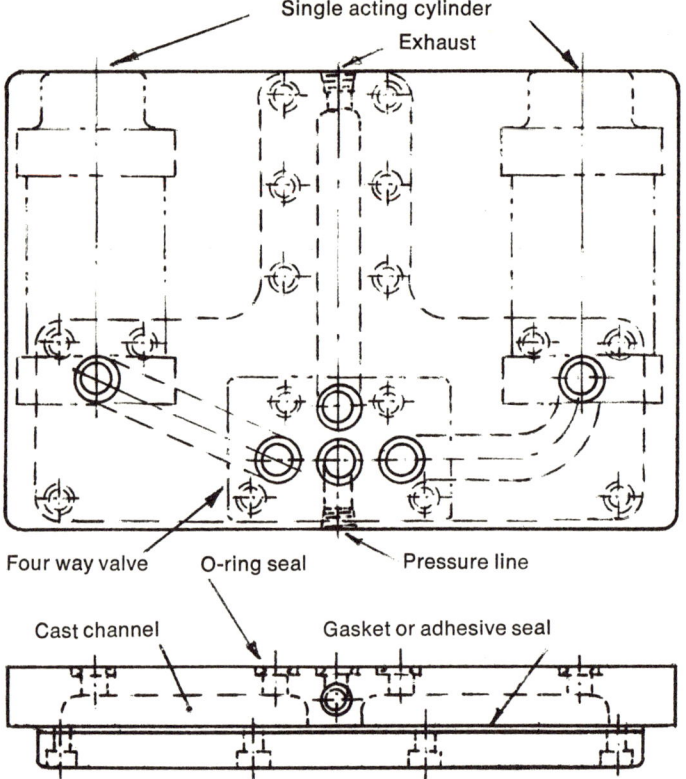

Figure 4.5-12. Manifold plate, cast construction

This design is for the same components as shown in Figure 4.5-11. It is a more direct approach with large radii possible for the channels. A straight channel from the valve to the cylinder on the left side of the drawing is the best. If, however, this would interfere with the fastening holes for the cylinder, a 90° channel with a large radius may be the best solution.

available from the pneumatic industry and competent representatives are willing to be of assistance. Some of the standard parts are:

1. Pneumatic chucks. A dependable holding device for rotating parts for lathes and other manufacturing machines where no hydraulic power is available. (Figures 4.5-13 and 4.5-14)
2. Pneumatic cylinders. These may in some cases be designed as an integral part of the machine. Many pneumatic manufacturers carry a number of standard sizes and shapes. (Figures 4.5-15 and 4.5-16)

4.5 Pneumatics

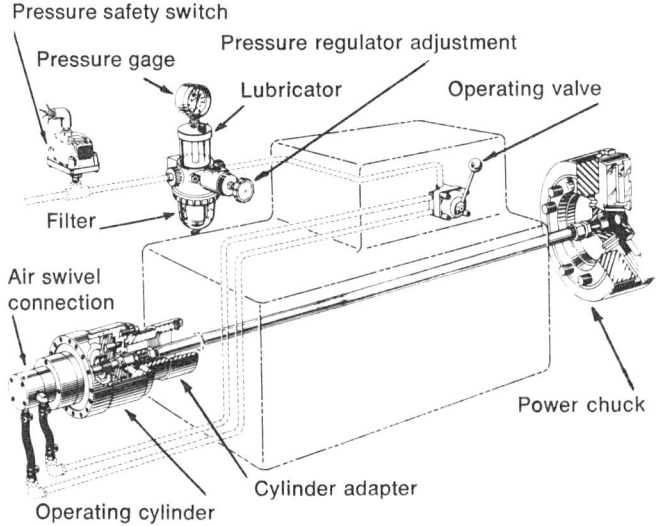

Figure 4.5–13. Typical pneumatic chuck installation for machine tool. (Courtesy of Cushman Industries, Inc.)

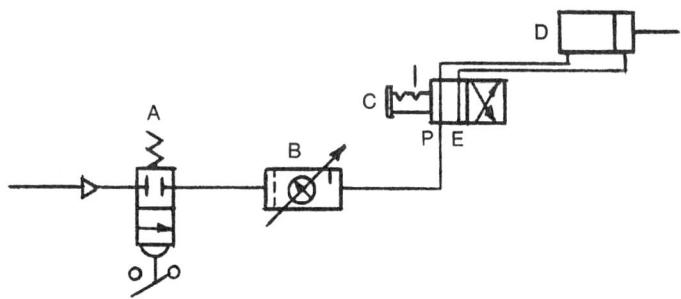

Figure 4.5–14. Pneumatic symbols for chuck layout
 A is a 2-way normally closed valve with a normally open electric switch. The electric switch closes the electric circuit when the air pressure is high enough for safe operation.
 B is a combination filter, regulator and lubricator with gage.
 C is a 2-position manually operated 4-way valve.
 D is the pneumatic chuck cylinder.

Prototype Machine

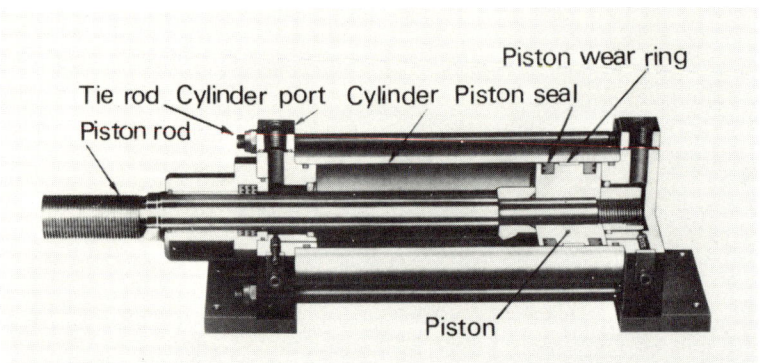

Figure 4.5–15. Typical pneumatic double acting cylinder. (Courtesy of Bellows-Valvair)

Table 4.5–16. Air Consumption and Developed Thrust

Cylinder Bore Diameter *	Volumetric Displacement	Air Consumption †	Theoretical Thrust in Pounds for Cylinder Bores and Pressures Shown				
			Air Line Pressure, psi				
Inches	Cubic ft.	Cubic ft.	40	60	80	90	100
1	.00045	.00183	31	47	63	71	78.5
1½	.00102	.00412	71	106	141	159	177
2	.00182	.00733	126	188	251	283	314
3	.00409	.01649	283	424	565	636	707
4	.00727	.02931	503	754	1005	1131	1257
6	.01636	.06595	1131	1696	2262	2545	2827
8	.02909	.11724	2011	3016	4021	4524	5026

* For advance stroke, rod under compression
† Free air at 90 psi for each inch of stroke

4.5 Pneumatics

When air enters a cylinder to perform work, the pressure rapidly builds up high enough to start the piston moving, and when the piston attains the end position, the pressure will reach the value set by the pressure regulator. This may be the ideal condition. If, however, the mechanism operated by the piston has a stick slip condition, speed control may have to be applied. Many cylinders are now designed to entrap the escaping air in such a way that bounce at the end of the stroke will be negligible. The cylinder diameter should be based on required load plus 50%, and is expressed thus:

$$D = \frac{1\frac{1}{2} F}{P_2}$$

$A =$ Cylinder area, $\frac{\pi}{4} \times D^2$

$F =$ Anticipated load

$P_2 =$ psi as set by the pressure regulator

For all practical purposes it may be assumed that the required pressure has been reached with no perceptible addition of heat, and air consumption may, therefore, be calculated on the theory of adiabatic compression. The volume is inversely proportional to the absolute pressure.

$$V_2 = \left(\frac{P_1}{P_2}\right)^{.71} \times V_1$$

It is most convenient to calculate the volume using psia and inches and divide the end result by 1728 to obtain cubic ft.

$P_1 =$ Initial absolute pressure, psia
$P_2 =$ Absolute pressure after compression, psia
$V_1 =$ Initial volume, cubic inches (Volumetric displacement)
$V_2 =$ Volume after compression of V_1
$k = 1.41$ for adiabatic compression $\quad \frac{1}{1.41} = .71$

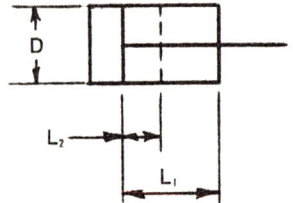

Figure 4.5–16a.

$P_1 = 14.7$ psia
$P_2 = 14.7 + 90 = 104.7$ psia
$V_1 = $ Area of cylinder bore $\times L_1$
$L_1 = $ Assume 1 inch stroke

$$V_2 = \left(\frac{P_1}{P_2}\right)^{.71} \times V_1$$

$V_1 = A \times L_1$

$V_2 = A \times L_2 \,;\, L_2 = \dfrac{V_2}{A}$

$$\frac{L_1 \times A \times L_1}{L_2} = \frac{L_1 \times A \times L_1 \times A}{V_2} = \frac{A}{\left(\dfrac{P_1}{P_2}\right)^{.71}} =$$

$$= \frac{A}{\left(\dfrac{14.7}{104.7}\right)^{.71}}, \text{ free air consumption, cubic inches per one inch stroke}$$

$\dfrac{A}{428.7} = $ free air consumption, cubic feet per inch of stroke

It should be noted that an additional amount of air is required for pipes, valves and cylinders at rest.

The following formula may be used for calculating the cycle speed:*

$$Q = .0273 \times \frac{D^2 \times L}{t} \times \frac{P_2 + 14.7}{14.7}$$

Q = Air flow in cubic feet per minute, free air
D = Cylinder bore, inches
L = Stroke, inches
t = Time to complete stroke, seconds
P_2 = Operating pressure psig

* Courtesy of C. A. Norgren Company.

4.5 Pneumatics

Other types of cylinders may also be designed for a great variety of purposes. Figures 4.5–17a and 4.5–17b show a simple design. Here a rolling diaphragm is used which effectively prevents leaks at the piston. In the one direction of stroke, the diaphragm rolls off the outside piston diameter onto the inside of the cylinder diameter. There is no friction involved, except as concerns the negligible amount of power necessary to flex the fabric. For best life, a fabric having free circumferential elongation and limited axial elongation is necessary. These diaphragms are available in standard sizes to fit many applica-

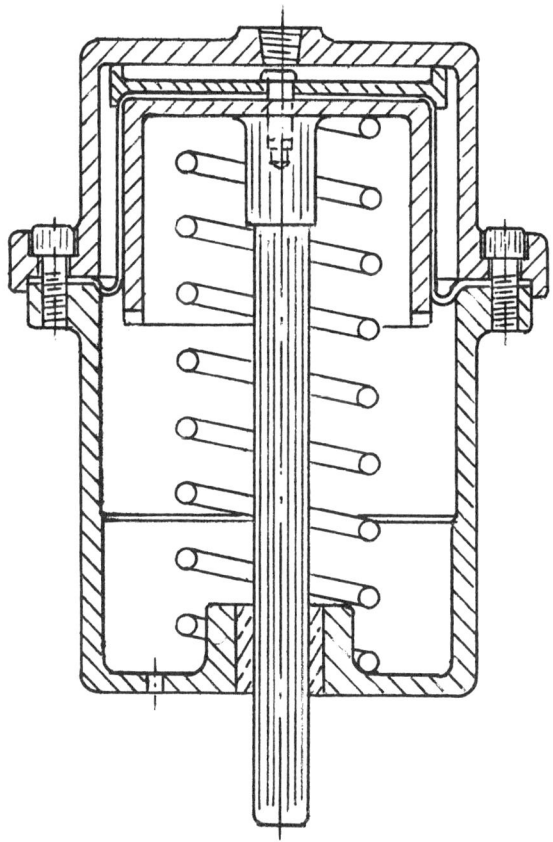

Figure 4.5–17a. (Permission to redraw by Bellofram Corporation)

Prototype Machine

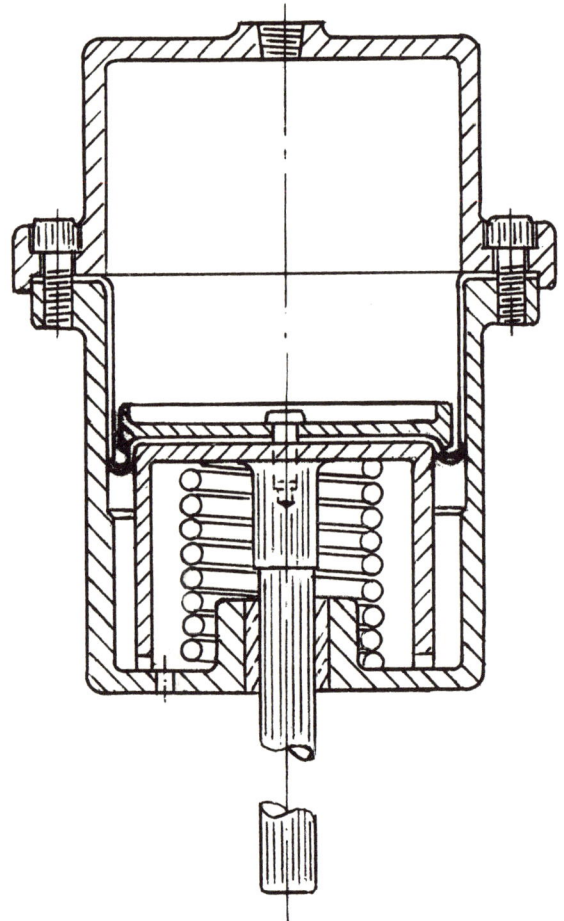

Figure 4.5–17b. (Permission to redraw by Bellofram Corporation)

tions. A flat diaphragm, properly arranged, is very useful for short motions.
3. Pneumatic valves. These components are designed to control, direct and stop the flow of air. They come in many different sizes and shapes to meet a great variety of conditions. They may be operated manually, mechanically by impulses from moving parts of the machine tool, electrically or pneumatically. Figures 4.5–18 to 4.5–23 are some of the accessory valves.

4.5 Pneumatics

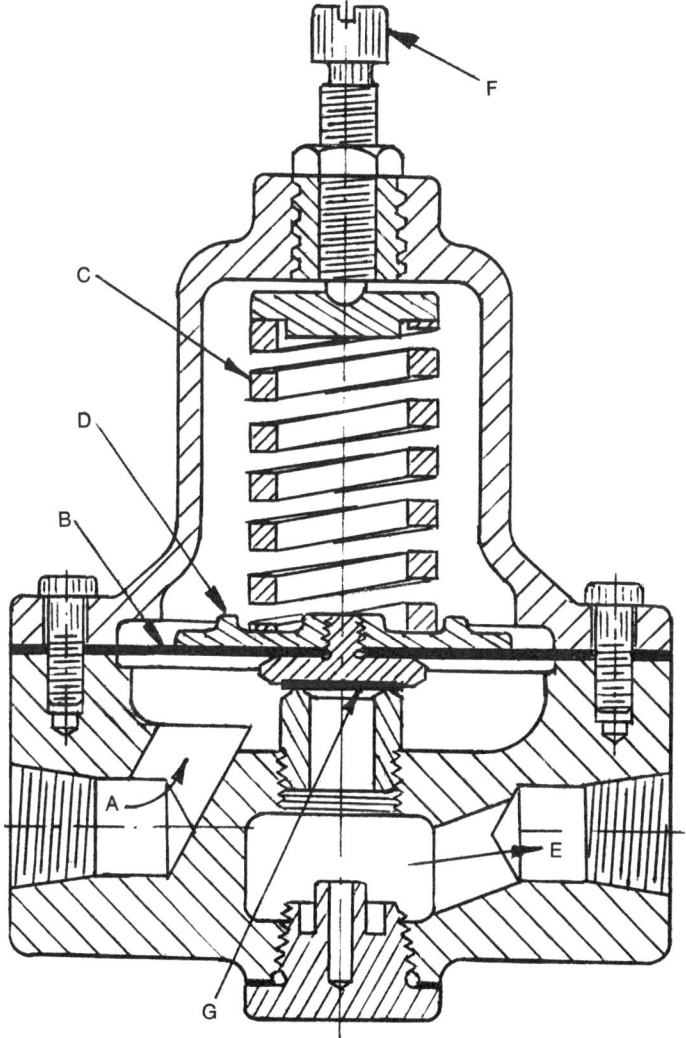

Figure 4.5–18. Diaphragm relief valve.

Primary air enters inlet port A, and flows through the valve body to exert pressure on the bottom of the diaphragm B. If the pressure exceeds the force exerted by the regulating spring C, the diaphragm moves upward with the spring rest D, allowing air pressure to escape through the valve to the secondary outlet port E. The desired relief pressure is set by the adjusting screw F.

When the excess pressure has been relieved, and the regulating spring force becomes equal to or greater than the system pressure, the seal, G, comes in contact with the seat, preventing further escape of pressure. (Courtesy of C. A. Norgren Company)

Prototype Machine

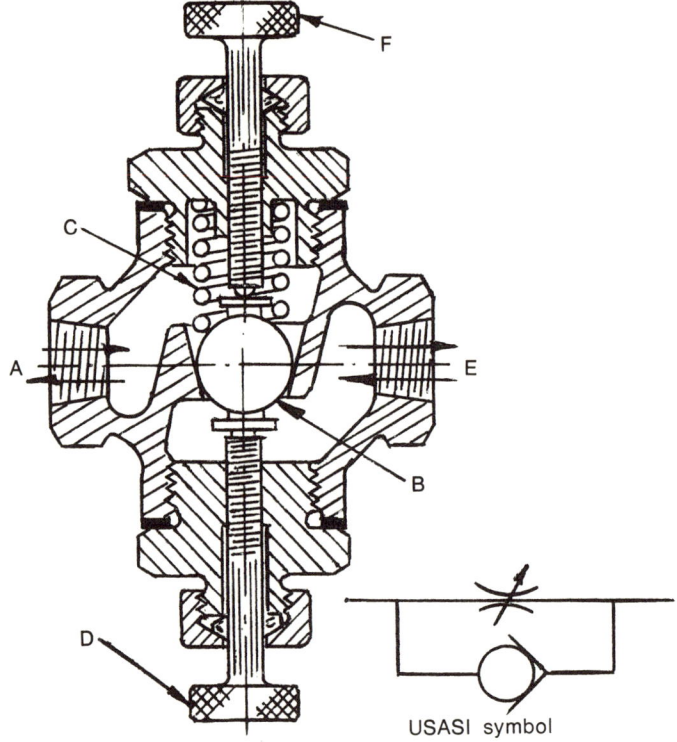

USASI symbol

Figure 4.5–19a. Flow control valve, two way.
Ball B is seated in a conical hole by spring C. By withdrawing adjusting screw F and advancing adjusting screw D very accurate flow control can be obtained, because the air must pass between ball B and conical seat when flowing from port A to port E or vice versa. (Courtesy of C. A. Norgren Company)

Figure 4.5–19b. Flow control valve.
Here air can flow freely in one direction and be metered through the valve in the opposite direction.
For free flow, the air enters port A, lifts the valve assembly B and flows unrestricted out through port E.
For restricted flow, the air enters port E and puts additional force on top of valve assembly B, which is already held against its seat with light spring pressure. The air can then flow only through adjustable orifice C and out through port A. The flow is controlled by adjusting screw D. (Courtesy of C. A. Norgren Company)

4.5 Pneumatics

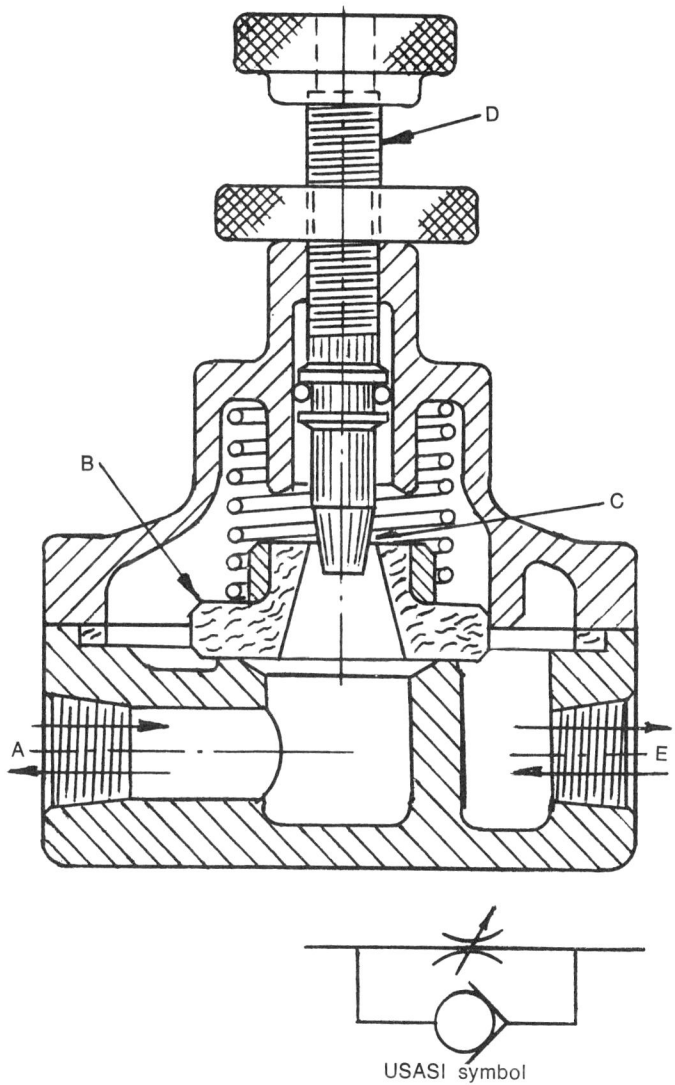

USASI symbol

Prototype Machine

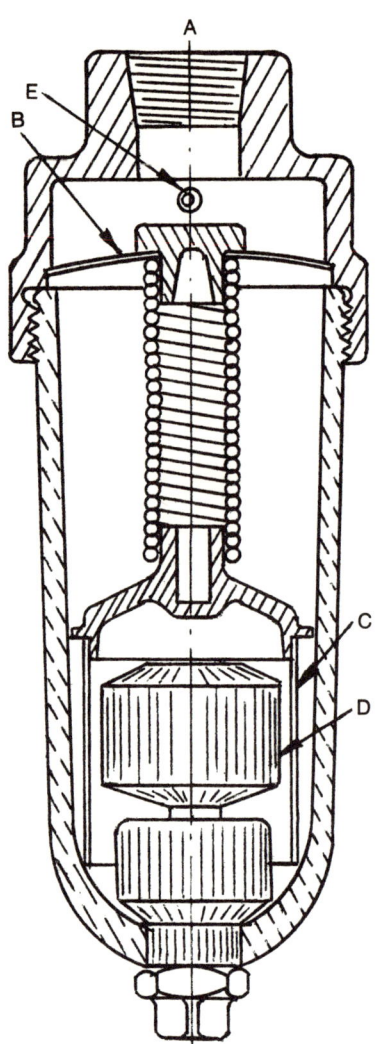

Figure 4.5-20. Drip leg automatic drain

Water accumulated in the air system enters at inlet A. Solid contaminants such as pipe scale, rust etc. are trapped by inlet screen B. The protective screen C further traps particles that will not freely pass through the drain valve. As water accumulates in the bowl, the float D will rise, opening the float valve. Bowl pressure is admitted to the area above the piston, allowing the piston to move downward, opening the drain valve and allowing the water to be expelled. The float valve will then close, allowing the drain valve to close.

During periods when the air line is not pressurized, the drain valve remains open, allowing water to drain by gravity. Approximately 10 psig system pressure is required to close the valve. E is vent valve. (Courtesy of C. A. Norgren Company)

4.5 Pneumatics

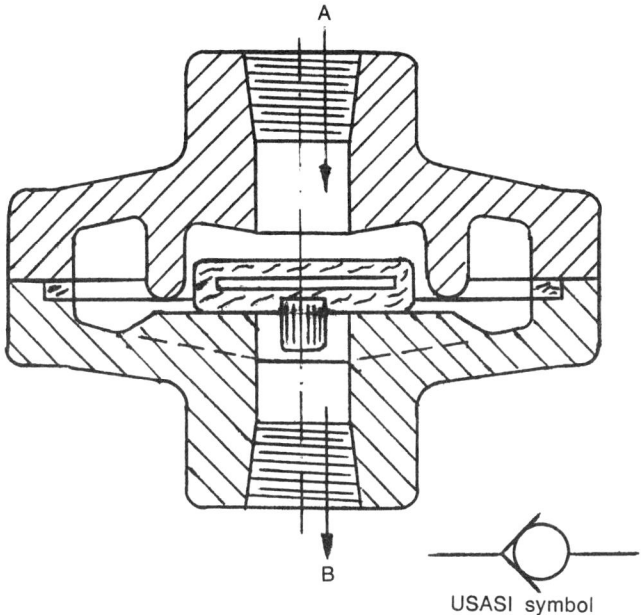

Figure 4.5-21. Check valve.
When inlet port A is pressurized, flow is permitted through outlet port B. When outlet port B is pressurized, flow is blocked by valve. (Courtesy of C. A. Norgren Company)

The most common air operators for control valves are used when it becomes necessary or desirable to control the operation of the main valve at some distance from the main valve. The air signal from the remote pilot valve impinges directly on the piston or pistons of the poppet valve to cause it to operate. When air is vented off through the remote pilot, the valve is returned to its normal position by the spring or springs in the valve and by the air load on the poppets.

OPERATION OF VALVES FROM A MAINTAINED AIR SIGNAL. In many installations the air operated valve receives a continuous or maintained air signal from a remote pilot valve. As long as this signal is maintained, the air operated valve will be operated. When the air signal stops, and is exhausted, the air operated valve returns to its normal position. Figure 4.5–24 illustrates this combination of 3-way pilot valve, 4-way air operated valve and cylinder.

OPERATION FROM MORE THAN ONE MAINTAINED AIR SIGNAL. In the

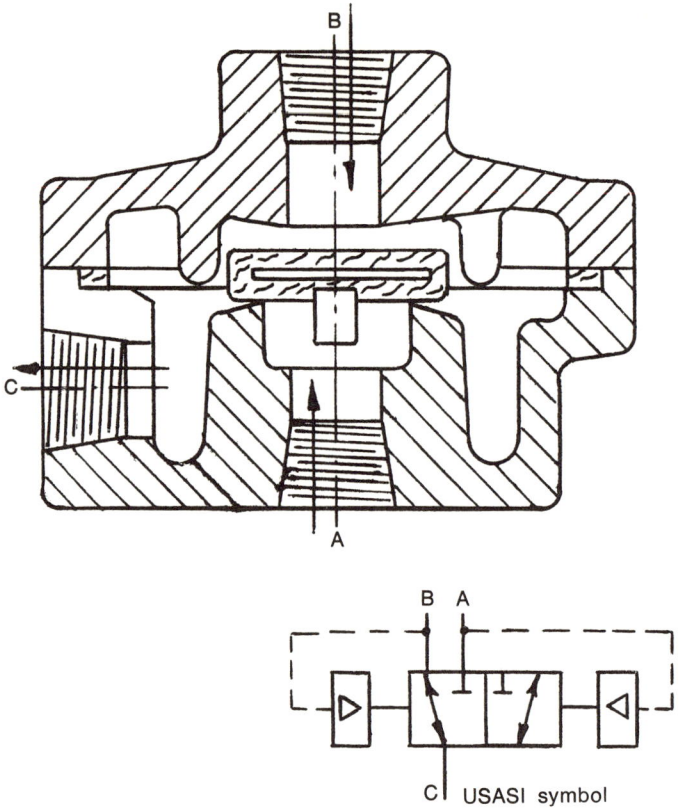

Figure 4.5-22. Shuttle valve.
When inlet port A is pressurized, flow is permitted through outlet port C. Inlet port B is blocked.
When inlet port B is pressurized, flow is permitted through outlet port C and inlet port A is blocked. (Courtesy of C. A. Norgren Company)

circuit shown in Figure 4.5-25, any one of three normally closed 3-way pilot valves will, when actuated, supply a signal to the air operated 4-way valve. When any one or any combination of the three pilot valves is actuated, the 4-way valve will be operated, and when all three of the pilot valves are released, the 4-way valve will return to normal position. Notice, that it is necessary in this case to employ two ¼ inch NPT shuttle valves in the circuit to prevent the inadvertent loss of a pilot signal from one pilot valve through the exhaust port of another pilot valve.

4.5 Pneumatics

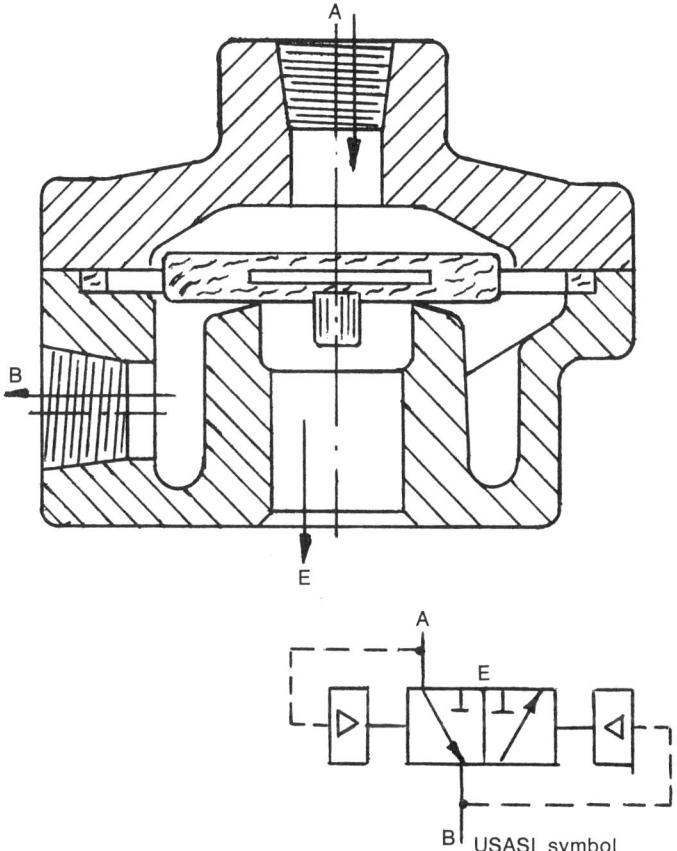

Figure 4.5-23. Quick exhaust valve.
When inlet port A is pressurized, flow is permitted through outlet port B. Exhaust port E is blocked.
When outlet port B is depressurized, flow is permitted through exhaust port E. Inlet port A is blocked. (Courtesy of C. A. Norgren Company)

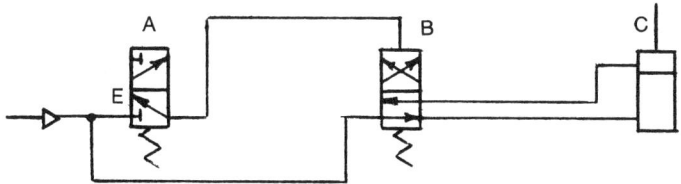

Figure 4.5-24. Operation of valves from a maintained air signal.
A is a 3-way pilot valve. B is a 4-way air operated valve and C is a cylinder. (Courtesy of C. A. Norgren Company)

Prototype Machine

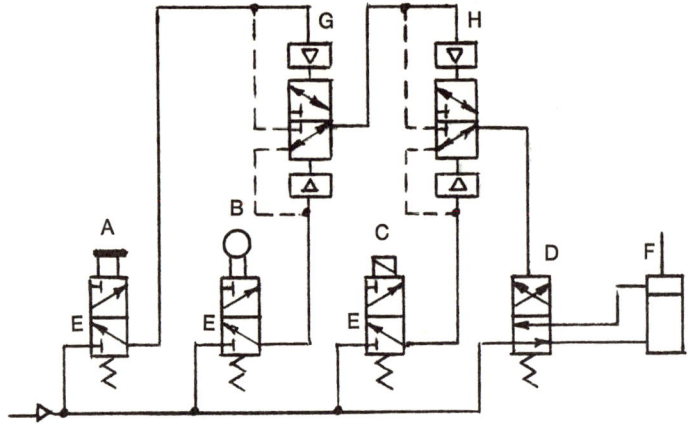

Figure 4.5–25. Operation from more than one maintained air signal.
Manually operated 3-way pilot valve A. Mechanically operated 3-way pilot valve B. Solenoid operated 3-way pilot valve C. Air operated 4-way valve D. Shuttle valves G and H and cylinder F. (Courtesy of C. A. Norgren Company)

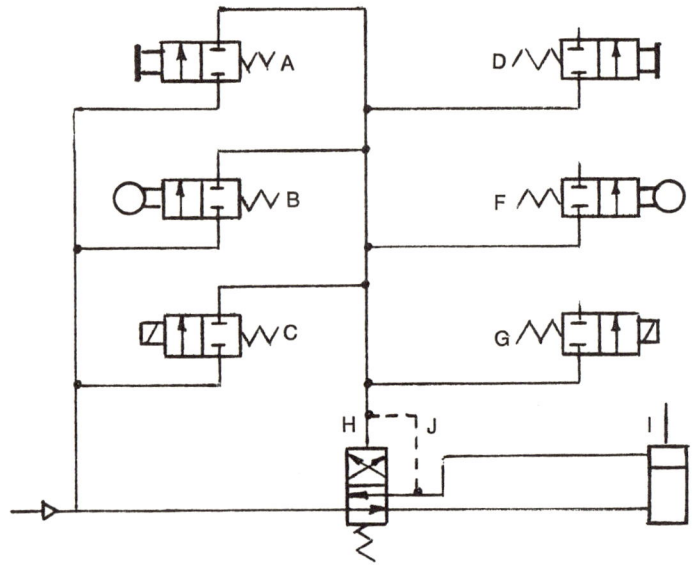

Figure 4.5–26. Operation from momentary pilot signal.
Three pressurizing 2-way normally closed pilot valves A, B and C. Three venting 2-way normally closed pilot valves D, F and G. Air operated 4-way valve H with built in sustained bleed J and cylinder I. (Courtesy of C. A. Norgren Company)

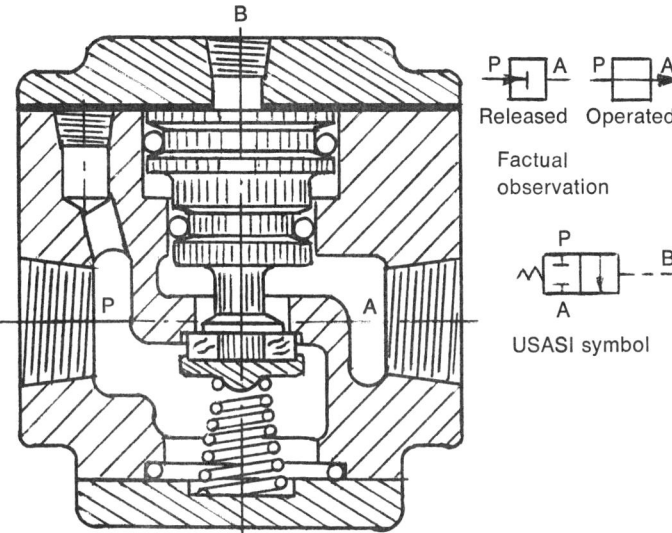

Figure 4.5–27a. Normally closed air operated 2-way valve.
For normally closed valve with pilot exhausted, the valve is closed. With pilot pressurized, the valve is open.
This valve is used for on and off control when it is necessary to locate the pilot valve some distance from the main valve. Pilot pressure is indicated at B. (Courtesy of C. A. Norgren Company)

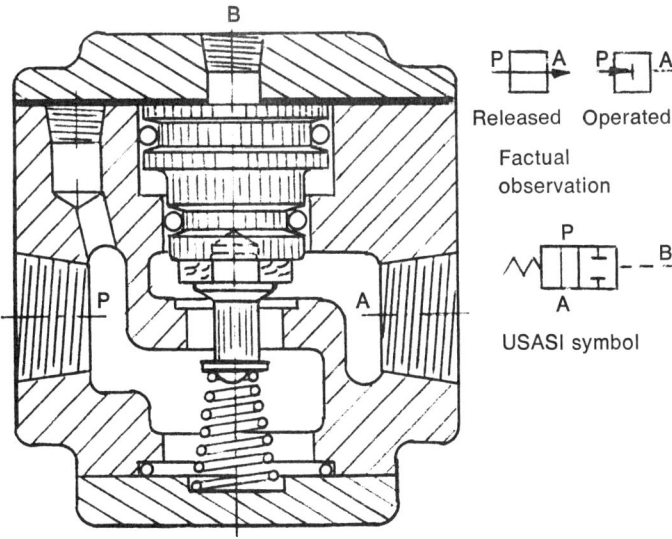

Figure 4.5–27b. Normally open air operated 2-way valve.
For normally open valve with pilot exhausted, the valve is open. With pilot pressurized, the valve is closed.
This valve is used for on and off control when it is necessary to locate the pilot valve some distance from the main valve. Pilot pressure is indicated at B. (Courtesy of C. A. Norgren Company)

Prototype Machine

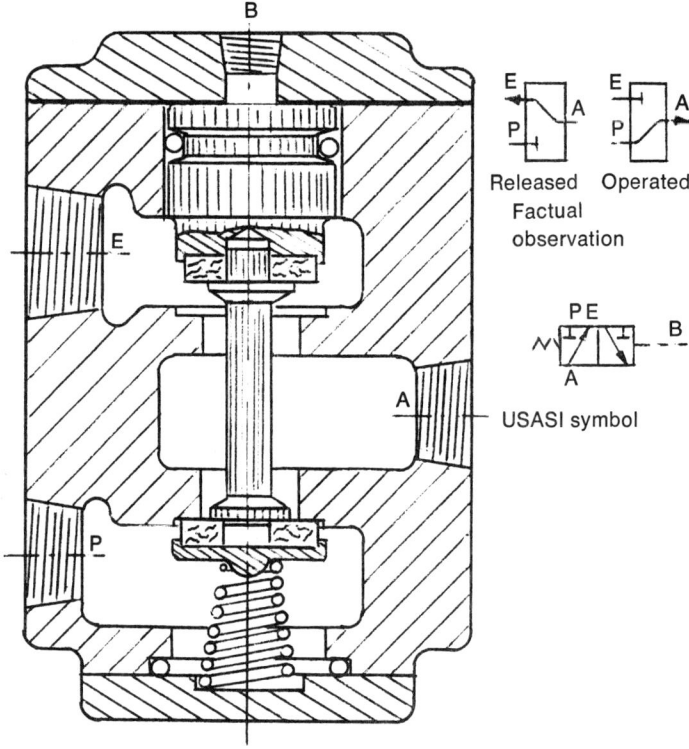

Figure 4.5–28a. Normally closed air operated 3-way valve.

For normally closed valve with pilot exhausted, outlet A is connected to exhaust E. With the pilot pressurized, outlet A is connected to pressure P.

This remotely actuated valve is used when it is necessary to locate the pilot some distance from the main valve. Pilot pressure is indicated at B. (Courtesy of C. A. Norgren Company)

OPERATION FROM MOMENTARY PILOT SIGNAL. Two-position, spring-returned valves are very well suited for momentary air operation. If the circuit shown in Figure 4.5–26 is examined, it will be noticed that the pilot operators indicated are normally closed 2-way valves, instead of the more expensive 3-way valves required for momentary operation of some spool valves. Shuttle valves are not required for momentary operation of these poppet type, two-position, spring-returned valves.

The 4-way valve will be operated when any one or more of the three pressurizing pilots are operated momentarily to admit air to the lines connected to the air operating head of the 4-way valve. When

4.5 Pneumatics

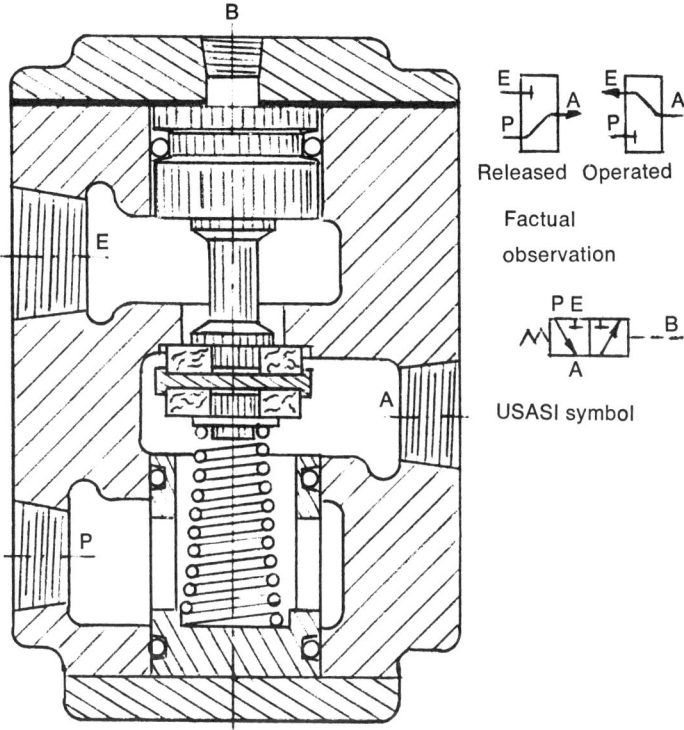

Figure 4.5-28b. Normally open air operated 3-way valve.
For normally open valve with pilot exhausted, outlet A is connected to pressure P. With the pilot pressurized, outlet A is connected to exhaust E.
This remotely actuated valve is used when it is necessary to locate the pilot some distance from the main valve. Pilot pressure is indicated at B. (Courtesy of C. A. Norgren Company)

the pressurizing pilot, or pilots again close, the air previously admitted to the 4-way pilot circuit remains trapped in that circuit, and the 4-way valve remains in the operated position until one of the venting pilots is operated momentarily to exhaust the trapped air and thus permit the 4-way valve to return to its normal position.

When this type of service is desired, the air operated valve must be purchased with a sustaining bleed, which is an internal connection from the normally closed cylinder port to the pilot chamber of the air operated valve. When the valve is operated, pressure to the normally closed cylinder port bleeds into the pilot chamber to offset any loss of

Prototype Machine

air from the pilot chamber due to minor leaks in the circuit. If the circuit is tight, no air flows in the bleed passage. There is no flow in the bleed passage when the valve is in the normal position, since the normally closed cylinder port is then open to the exhaust.

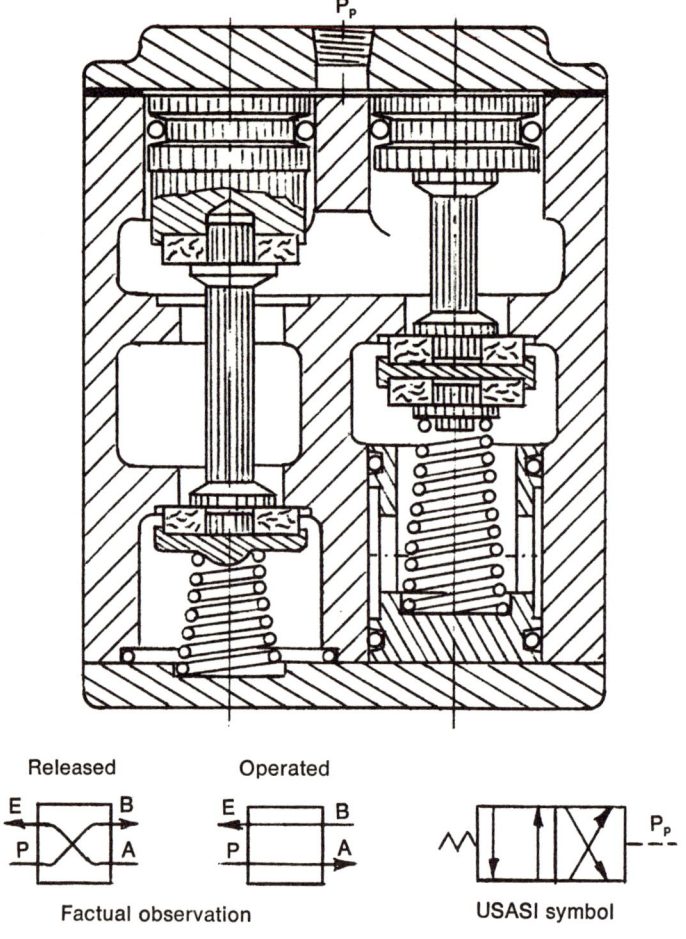

Figure 4.5–29. Air operated 4-way valve.
   When pilot is exhausted, cylinder port A is exhausted and cylinder port B is pressurized.
   When pilot is pressurized, cylinder port A is pressurized and cylinder port B is exhausted.
   This valve is used to control double acting cylinders where it is necessary to locate the pilot valve some distance from the main valve. Pilot pressure is indicated at P_p. (Courtesy of C. A. Norgren Company)

4.6 Hydraulics

The possibilities for interesting and useful applications of air circuitry are many. In Figure 4.5–26, three pressurizing pilots and three venting pilots have been used as illustrations. Any number of pressurizing or venting pilots, from one up, may be employed as required to accomplish the desired valve operation.

Figures 4.5–27 to 4.5–29 illustrate some of the most common valves.

Quite often it becomes necessary to design your own valves and cylinders. The design of a certain machine part may make this desirable, or maybe the space available will prompt this decision. A spool design as shown in Figure 4.5–30 may then be the answer.

Sometimes, in an automatic machine, it is advantageous to operate the control valves mechanically from the motions in the machine drive. A dependable arrangement for mechanical control is a number of cams arranged on the same shaft which can be driven at a constant speed with a synchronous motor. (Figure 4.4–6) Adjustments and variations can be provided for by the cams. The synchronous motor can be started with a limit switch tripped by a cam in the main machine drive, and stopped by a limit switch tripped by a cam driven by the synchronous motor.

4.6 Hydraulics in Machine Tool Design

Hydraulics, as well as pneumatics, play an important part as auxiliary equipment for machine tools. The fluid used is almost universally oil. If properly selected, it has good lubricating qualities, which is essential where materials subject to corrosion are used.

In some cases, where a fire hazard might exist, the flash point of the oil should be considered. At room temperature, the ordinary hydraulic oil will burn, if a wick of some kind, immersed in the oil, is ignited with an open flame.

The flash point of the oil is the temperature to which the oil must be heated to give off enough vapor to ignite and burn continuously without the aid of a wick, when an open flame is applied to it. The flash point of most petroleum oils, which are good hydraulic oils for machine tools, lies between 350 to 450°f.

The spontaneous ignition temperature of petroleum oils used for machine tools is the temperature at which the oil will ignite without outside help for ignition, and ranges between 450 and 650°f.

The best operating temperatures for a hydraulic system for ma-

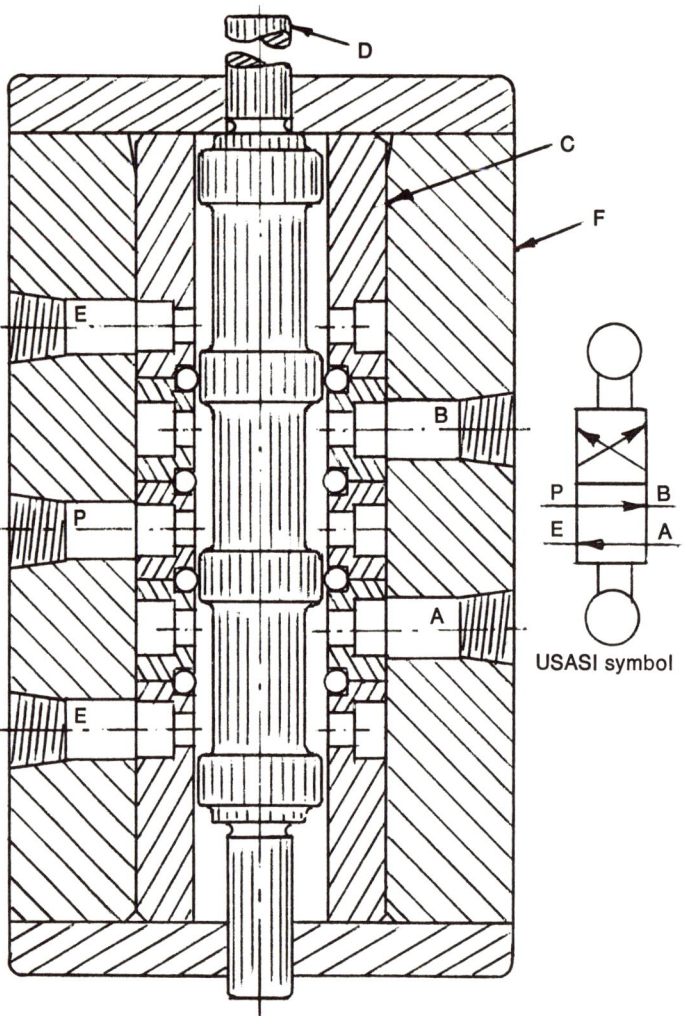

Figure 4.5–30. Mechanically operated 4-way valve.

This drawing shows a 2-position spool valve which may control the motions of a double acting cylinder. The pressure P stays in the central spool groove for both positions of the valve. As shown, the pressure P leaves port B for one end of the cylinder. The other end of the cylinder is connected to port A and is open to exhaust port E.

If the valve is shifted to the opposite position, the pressure P is open to the cylinder through port A and port B is open to exhaust E.

Spool D may be operated manually, mechanically, by air or even directly with a solenoid. If a good supply of lubricant is available, the spool clearance may be a honed fit and there would then be no need for a packing. If, however, you must depend on the lubricant carried in the air supply, it may be necessary to provide for packing. Many types of packing are available. An O-ring is shown in the illustration, but this could also be a square packing, because the thick-

4.6 Hydraulics

chine tools is in the neighborhood of 120°f. When a high degree of accuracy is required from the hydraulic system, the oil temperature should be kept uniform, because the viscosity of the oil varies quite a bit with a change in temperature, resulting in a change in operation. To accomplish this, it may be necessary to use a heat exchanger to get rid of the heat.

All engineering materials, when under pressure, will compress. If the pressure on oil is high enough, the magnitude of compression is measurable. The oil is also confined by other engineering materials that are subject to elastic displacements.

The hydraulic pressure used for machine tools varies in most cases from 200 to 1000 psi; but higher pressures may also be used. The higher pressures present more of a problem in sealing, loss of oil, and interrupted production. The average hydraulic pressure of about 600 psi is much higher than the common pneumatic pressure of 80 to 90 psi. Accordingly, a hydraulic chuck, for instance, is much smaller than a pneumatic chuck doing the same work. The space saving advantages of the smaller hydraulic components are very apparent.

Before going into the application of hydraulics to machine tools, it is necessary to understand the operation of an efficient hydraulic system in the simplest form. Four main components are required, all made in several sizes and with many different operating characteristics.

1. The pump, as a rule electric motor driven, supplies kinetic energy to the operating components. (Figure 4.6–1) The hydraulic oil is almost always contained in a separate tank set next to the machine tool. This way, it is much easier to keep the oil at the desired temperature than if the oil were kept in

---

ness can be arranged so that the diameter of the seal would be squeezed down as the sleeve assembly C is put in place.

Even without packing, the inclusion of sleeve C in the design is the best arrangement since the spacing of the small holes in the sleeve connected to the grooves in the spool can be manufactured to a high degree of accuracy. Because the sleeve C also has annular grooves over the holes, the ports may be arranged in many different locations, and the accuracy of the valve is not dependent on the spacing of the port holes. If the valve is richly lubricated, the exhaust ports E may be connected in the housing F and the excess lubricant expelled through the exhaust may be directed to a place where it could be reused and would not do any harm to material being processed in the machine.

This is the basic design of a simple valve. Many variations are possible for solving a number of interesting problems.

Figure 4.6–1. Hydraulic pump. (Courtesy of Vickers, Inc.)

Figure 4.6–2. Hydraulic power unit. (Courtesy of Vickers, Inc.)

4.6 Hydraulics

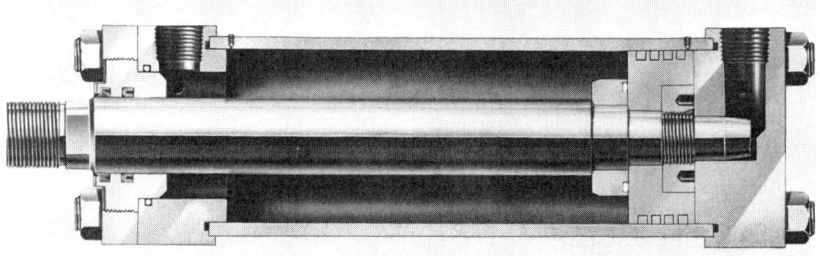

Figure 4.6–3. Hydraulic cylinder. (Courtesy of Vickers, Inc.)

the machine frame. The pump is usually mounted on top of the tank cover. (Figure 4.6–2)
2. The actuator may be a linear motion cylinder and piston, or it may employ a rotary motion. The linear motion actuator (Figure 4.6–3) is the most common, because most often rotary motion can best be performed with other mechanical means.
3. The valves direct, control, start and stop the flow to and from

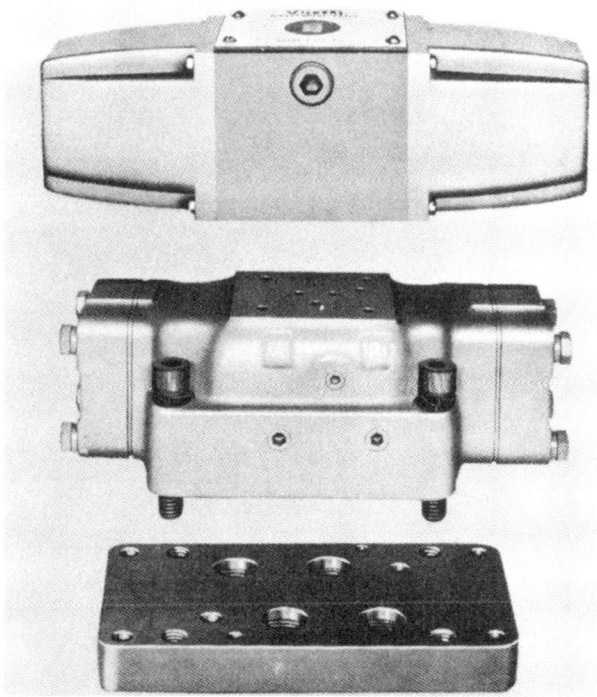

Figure 4.6–4. Sub-base mounted hydraulic valve. (Courtesy of Vickers, Inc.)

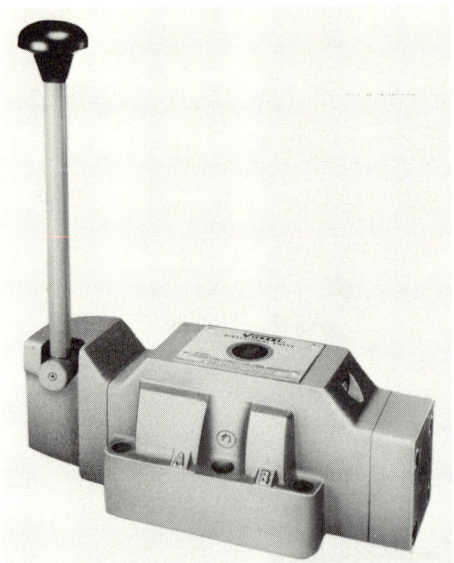

Figure 4.6–5. Lever operated directional hydraulic control valve. (Courtesy of Vickers, Inc.)

Figure 4.6–6. Knob operated directional hydraulic control valve. (Courtesy of Vickers, Inc.)

the components. The best valve for machine tools is a sub base mounted valve that may be easily replaced without disturbing the hydraulic pipes. (Figures 4.6–4, 4.6–5 and 4.6–6) If several valves can be placed close together, the best solution is a manifold plate. The holes from valve to valve and sometimes even to cylinders may be drilled or cast in the manifold plate. (Figures 4.5–11 and 4.5–12)

4. Chucks and holding devices are used for holding or clamping a work piece securely. (Figure 4.6–7)

4.6 Hydraulics

Accumulators, heat exchangers and oil filters must be considered as accessories and come in many different sizes and shapes, and with varieties of operating characteristics.

Accumulators store hydraulic energy, usually, for machine tools, by compressing a gas. They have many uses, the most common being safety measures.

   a. Where a pulsating load, an abruptly applied load or increasing back pressure occurs, the accumulator will protect the system from injurious results of increased pressure.
   b. When power failure occurs, or if the machine tool for some reason is stopped, the accumulator will hold a work piece, secured with a hydraulic chuck or clamp, in the proper working position until the power is restored.

Heat exchangers are used for cooling the hydraulic fluid when accuracy calls for an even temperature of the fluid.

Filters and strainers are important accessories for a hydraulic sys-

Figure 4.6–7. Hydraulic chuck. (Courtesy of Gleason Works)

Prototype Machine

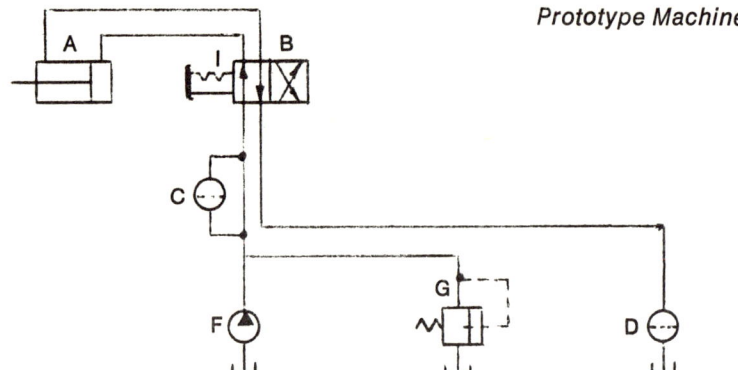

Figure 4.6–8. Simple hydraulic system.

The hydraulic oil must be filtered to get rid of undesirable contaminants. This is especially true when applied to a high grade machine tool. The filtering starts at the tank which should have a fine screen at the inlet when replenishing the oil. The tank must be thoroughly covered to prevent contaminants from entering from the air. The tank must have a breather hole, since the level of the oil varies through the machine cycle. This air intake must also be screened.

If one filter is used for the fluid lines, it may be placed in the pressure line or return line. Both methods have their advantages and disadvantages. A pressure line filter may cause too much pressure drop and should therefore have a bypass valve for safety. One way as shown at C is to connect the filter in parallel to the pressure line. The filtering is then, however, not as efficient. If the filter is put in the return line, as shown at D, back pressure may be a problem. Some conditions may require two filters in the system.

Pump F circulates the fluid through hand operated 2-position 4-way valve B to operate double acting cylinder A. Relief valve G controls the system pressure.

Figure 4.6–9a. Electrically operated relief valve. (Courtesy of Vickers, Inc.)

4.6 Hydraulics

tem. Their function is to remove impurities absorbed as the fluid circulates, and to prevent them from reaching important working parts of the system.

A simple hydraulic system is shown in Figure 4.6–8. This shows a relief valve which is set sufficiently high to carry the work load. When the piston reaches the end of the stroke, the pressure builds up to the value set at the relief valve and the oil returns from the pump to the tank through the relief valve. The piston is then firmly held in position with the pressure set at the relief valve. (Figures 4.6–9a and 9b)

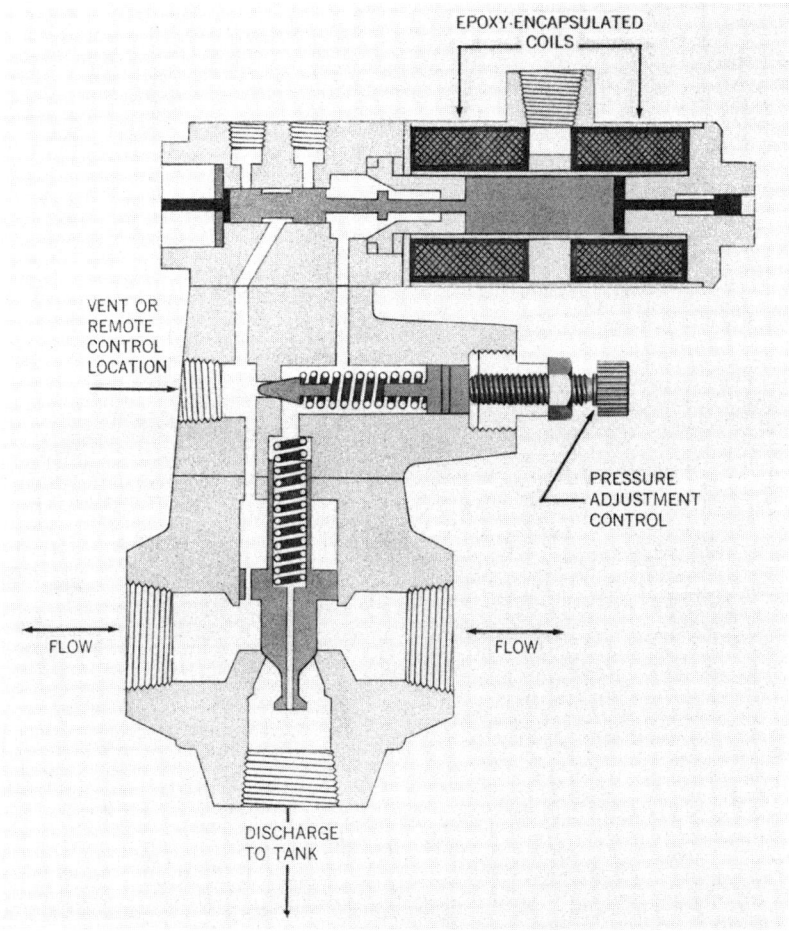

Figure 4.6–9b. Sectional view of electrically operated relief valve. (Courtesy of Vickers, Inc.)

173

Prototype Machine

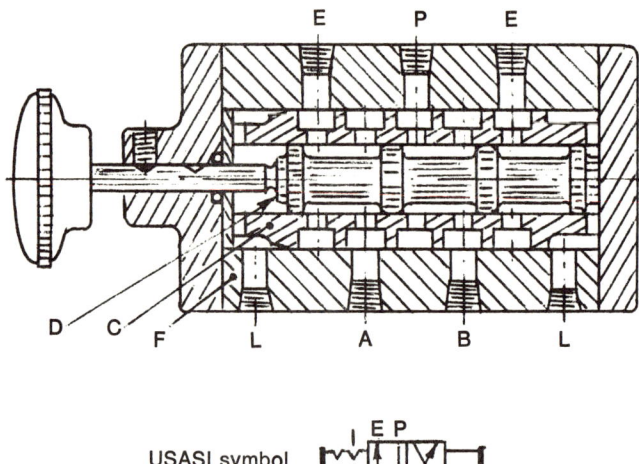

Figure 4.6–10. Hand operated 2-position 4-way valve.

No air should be trapped in the valve, for this would hamper the operation. Therefore, the sections at the ends of the spool D are vented through L. Since some oil will leak through, the leak returns L should be piped to the sump so no oil will escape to places where it could do damage.

Pressure enters the valve at P and in the position shown leaves port B to operate one end of a double acting cylinder. The other end of the cylinder is on exhaust through port A.

Pulling the spool D out reverses the cylinder stroke. Sleeve C makes it easy to make the valve integral with some other machine member. This design also makes sub-base mounting easy because holes can be on one side of body F. The grooves in sleeve C permit ample spacing of all holes.

The simplest form of hydraulic valve is shown in Figure 4.6–10. This is a hand-operated 2-position 4-way valve, a basic design easy to incorporate in a machine member if it is necessary or desirable to design your own valves and cylinders. The cast iron sleeve makes it possible to locate accurately the holes leading to the grooves in the spool, using standard manufacturing procedures. Be sure that the holes and grooves are of sufficient area to carry the volume required for best performance.

Entrapped air in the hydraulic fluid may cause trouble, impairing the accuracy required. It is advisable to provide bleeder holes in the upper parts of the hydraulic system, an invaluable aid when first starting the machine, or when air has accumulated.

4.6 Hydraulics

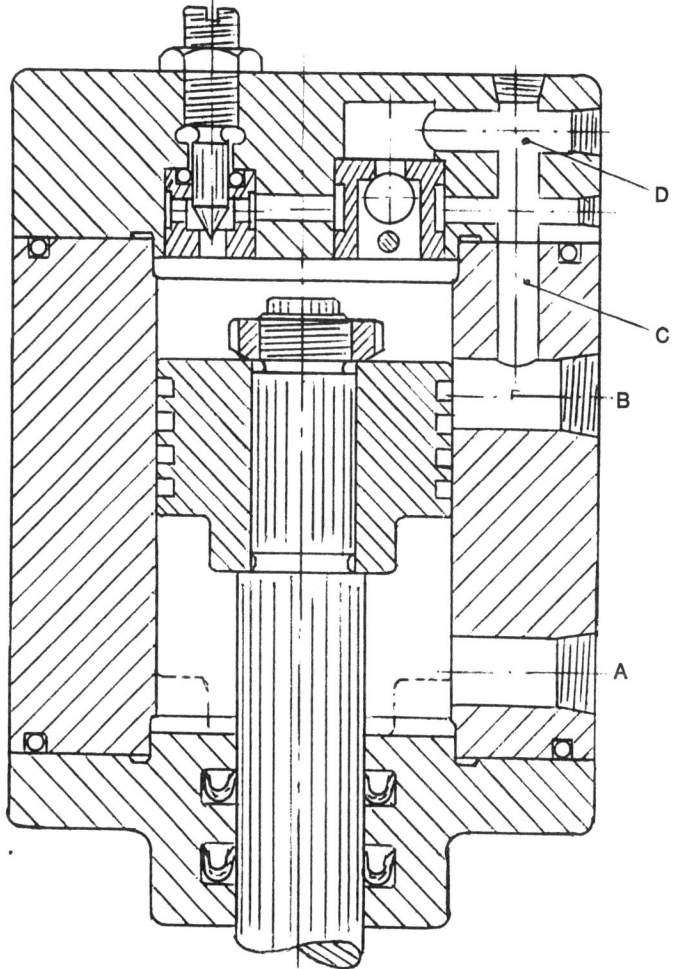

Figure 4.6–11. Metering valve.

When port B is pressurized, the oil will travel through holes C and D to open the ball check valve. Full flow starts as soon as cylinder port B is open direct to the cylinder.

When port A is pressurized, the oil enters the cylinder direct to push the piston to the right. Port B is then open direct to exhaust and partial pressure from the exhaust closes the ball check valve. As soon as port hole B is completely closed, the exhaust can escape only through the metering valve to port B.

This type of metering can also be applied to port A.

Prototype Machine

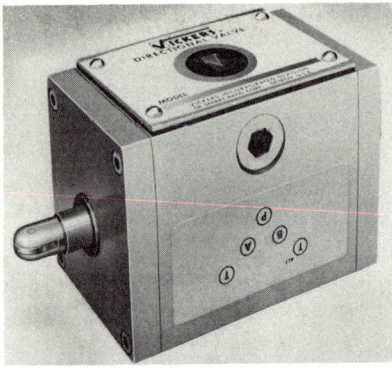

Figure 4.6–12. Cam operated deceleration valve. (Courtesy of Vickers, Inc.)

Speed control is necessary for heavy, hydraulically operated machine slides, usually at both ends of the stroke. This is done with metering valves, and the best results are obtained by metering out, when cushioning the end of the stroke; that is, arranging for the exhaust to be restricted as soon as a slowdown is desired. (Figure 4.6–11) This is a simple design, metering out through an adjustable metering valve, also called needle valve. In this case, the metering valve and check valve are both contained in the cylinder cap. This type of meter-

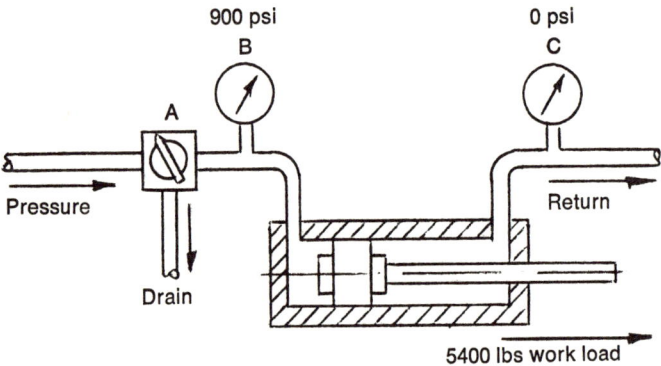

Figure 4.6–13a. Meter in control.
Schematic diagram showing reaction of work load fluid pressure in a double acting cylinder having full piston area and piston annulus in the ratio of 6:5 square inches.

This control system is recommended for feeding grinder tables, welding machines, milling machines, and rotary hydraulic motor drives. The flow control is adjusted with valve A shown here and in figures 4.6–14a and 4.6–14b. The line pressures are indicated by gages B and C. (Courtesy of Vickers, Inc.)

4.6 Hydraulics

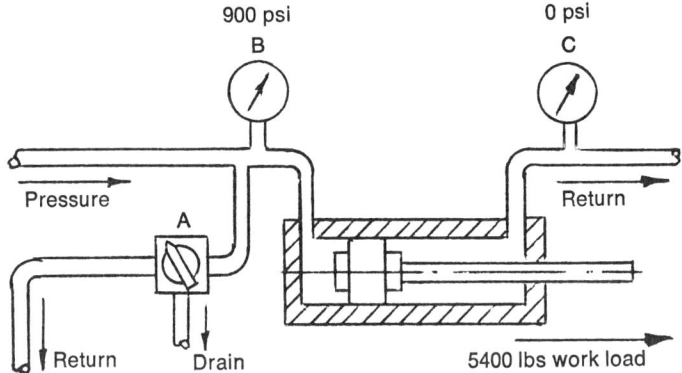

Figure 4.6–13b. Bleed off control.

Schematic diagram showing reaction of work load on fluid pressure in a double acting cylinder having full piston area and piston annulus in the ratio of 6:5 square inches.

This control system is recommended for reciprocating grinder tables, broaching machines, honing machines, and rotary hydraulic motor drives. The flow control is adjusted with valve A shown here and in figures 4.6–14a and 4.6–14b.

The line pressures are indicated by gages B and C. (Courtesy of Vickers, Inc.)

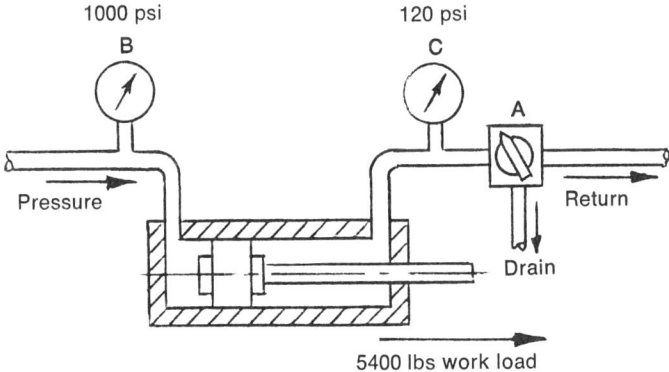

Figure 4.6–13c. Meter out control.

Schematic diagram showing reaction of work load on fluid pressure in a double acting cylinder having full piston area and piston annulus in ratio of 6:5 square inches.

This control system is recommended for drilling, reaming, boring, turning, threading, tapping, cut-off, and cold sawing machines.

The flow control is adjusted with valve A shown here and in figures 4.6–14a and 4.6–14b. The line pressures are indicated by gages B and C. (Courtesy of Vickers, Inc.)

177

Prototype Machine

ing functions equally well with a stationary cylinder or a stationary piston.

There are many cases when speed control is necessary during different portions of the manufacturing cycle of a machine tool. Then a cam operated valve may be the best arrangement. (Figure 4.6–12)

Some operations require meter in control, shown in Figure 4.6–13a; other operations require bleed off control, shown in Figure 4.6–13b; and still other operations require meter out control as shown in Figure 4.6–13c. It should be noted that the minimum practical metering rate required quite often determines the size of the cylinder, especially when slow rates of motion are required. A flow control valve is shown in Figures 4.6–14a and 14b.

Very complicated, irregular shapes may be machined with the use of hydraulic tracers. The author has had cases where a hydraulic tracer (Figure 4.6–15) has been used to a high degree of accuracy, guiding a small grinding wheel on an irregular, interrupted contour. Here, the pressure of the grinding operation was very low. A low hydraulic pressure could then be used and a stiff mechanical structure could be designed around that part of the hydraulic system to insure accuracy. Since the grinding wheel had to descend several inches vertically at a rapid rate of speed, it overshot a couple of thousandths of an inch when the stylus first made contact; but then corrected its position before the grinding wheel made contact. This operation called for a good grade of hydraulic oil, specified by the manufacturer of the tracer. To stabilize the action, a small back pressure was used in the exhaust line.

Figure 4.6–14a. Flow control valve. (Courtesy of Vickers, Inc.)

4.6 Hydraulics

Cams are often used in machine tools for performing a predetermined motion. When moving a light member, there is usually no problem. Because a small roller or cam rider is contacting the cam, the unit pressure is sometimes very high when moving a heavy slide, sometimes under a heavy load from the tools. This unit cam pressure, may be made very small by letting a hydraulic tracer command the motion of a heavy slide through a hydraulic cylinder and piston.

Another use of hydraulic power is for dressing a grinding wheel. By metering the oil flow, the speed of the cylinder or piston carrying the diamond across the grinding wheel can be very accurately controlled.

If hydraulic power is not available on the machine tool, hydraulic dressing can be very simply performed with a closed hydraulic system. A spring on the piston rod can be loaded up either by hand or by the motion of a machine member. When the loaded spring is released, it will push the piston against hydraulic fluid, forcing the fluid through an adjustable metering valve. When the spring is being loaded up, a check valve will let the oil follow the moving piston. This check valve will then close as the oil is being metered out through the metering valve. This will work well for a short distance. For a wide grinding

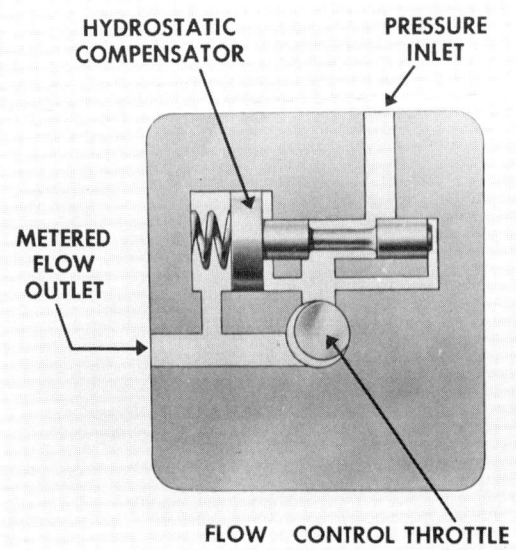

Figure 4.6–14b. Sectional view of flow control valve. (Courtesy of Vickers, Inc.)

Figure 4.6–15. Hydraulic tracer application. (Courtesy of Gleason Works)

wheel, air may be used as power supply instead of a spring. Air is, however, not dependable for accurate speed control.

A closed hydraulic system requires a small amount of liquid and can, as a rule, be very well protected from contamination.

Hydraulic power has long been used for presses, especially where much power is required.

4.7 Material Selection and Heat Treatment

There are many other hydraulic applications which an ingenious engineer will readily recognize. This chapter has presented some of the basic problems.

4.7 Material Selection and Heat Treatment in Machine Tool Design

In many cases, two or several different materials are under consideration for a machine member. Quite often, one will function just as well as the other, and the only deciding factor, for a wise selection, may be cost. The design engineer should exercise good judgment and choose the material best suited to fill all requirements.

For simplification, we will group the various components as follows:

1. Housings and Frames
2. Worms, Worm Gears and Gears in General
3. Shafts, Spindles, Plungers and Rods
4. Levers, Connecting Links and Minor Parts
5. Cams, Rollers and Riders
6. Moving Members in Contact where No Lubrication Is Available

Housings and Frames

The most common materials for these machine members are cast iron, and fabricated steel (steel weldments).

If the main functions of these parts are to support other machine members and to act as enclosures for lubricants, the selection may be based entirely on material and manufacturing cost.

If the design is such that few core boxes and loose pieces are required for the pattern, and the production rate is high enough, cast iron may be the answer since machining and material costs are low.

If the construction is simple so that the ribs and supports are easily accessible for welding, and especially if only one or two parts are required, fabricated steel should be considered. If machining must be performed on the parts after welding, the surfaces to be machined should first be rough finished and then the structure should be normalized or stress relieved in a furnace before the final finishing operation. This additional cost must be considered before finally deciding on which material to use.

There may be other considerations which alone would determine the choice of material.

SLIDING SURFACES. Cast iron sliding surfaces are in most cases satisfactory for housings and frames when loads are moderate and speeds are low. Cast iron has excellent wear qualities in contact with cast iron or soft steel. Cast iron must, however, always be well lubricated where motion occurs.

VIBRATION. For some machine tools it is of great importance to consider the effects that vibration will have on the parts being processed, when selecting the material for housings and frames; or if this is not detrimental, to consider the effects on other parts of the machine or adjacent machines. It is then well to consider the following:

1. Removal of Cause at Origin
   a. Preventing unbalance, if possible.
   b. Preparing proper support for tools and parts being processed.
   c. Using good judgment in selecting and preparing materials and supports for machine components, such as slides, shafts, rods, bearings, gears, cams.
   d. When vibration is of the resonant type as in a shaft, the shaft can be designed and supported, in most cases, so that the critical speed of the shaft is well above the operating speed. In cases of high speed machinery, the shaft and its mounting may have to be designed for the shaft to operate between the first and the second critical speeds.
2. Isolation or Preclusion of Vibration.
   The machine may be in such a category that vibration originating in the machine, either from machine operating components or from parts being processed, has no deteriorating effect on the parts being processed. This vibration may, however, have an undesirable effect on adjacent machines. In such cases, the vibration should be isolated by elastic suspension methods.
3. Absorption or Damping of Vibration.
   Resonant vibration in machine tools may be reduced or eliminated by these methods. It is well known that cast iron piston rings, successfully applied to steam engines and internal combustion engines from the beginning, were selected because they possessed good resiliency with accurate repeat through a long life, using grey cast iron as cast. Cast iron will return to its original state as soon as the vibration has dissipated in the heavier parts of the structure. The supporting ribs in a machine

4.7 Material Selection and Heat Treatment

housing or frame should, therefore, be adequate in number and of a uniform thin section, close to the load. The author has actually observed a case in which a machine that had been properly designed for dissipating vibrations and that had worked successfully for years was redesigned by an individual having limited practical experience. Unusually heavy ribs were provided in the housing supporting the tools and it had the effect of vibration marks on the parts being processed which never occurred on the previous design, and which disappeared when the design was corrected using more and thinner ribs.

Low carbon steel, which is used for fabricated structures, does not possess the quality of good resiliency, as cast iron does. This becomes an important factor when selecting the material for a machine tool housing or frame.

HEAT DISSIPATION. Sometimes it is important to depend on the machine housing or frame to dissipate heat generated from the processing of parts. Because the coefficients of heat transmission and heat radiation are considerably higher for cast iron than for steel, cast iron would be a wise selection.

These requirements must then be considered when deciding between cast iron and fabricated steel on the basis of cost.

Worms, Worm Gears and Gears in General

Soft steel worm and cast iron worm gear is an excellent combination for hand operated adjustments or power drives at moderate speeds and moderate loads. The lubrication must be proper for the best wear of the cast iron. For greater accuracy, a case hardened and ground alloy steel worm mating with a cast iron worm gear is recommended, and the worm should preferably be integral with the shaft.

For higher speeds, a worm of this material and preparation should mate with a bronze worm gear, well lubricated.

For gears in general, a good combination for hand operated adjustments and slow speed power drives of minor importance is a soft steel pinion and cast iron gear well lubricated. Most power drives should have a gear combination of a good grade of case hardened and ground alloy steel for the higher speeds, assuring accuracy and quietness. Sometimes a honed or lapped combination will add to these qualities. A hunting tooth combination is then recommended, where the number of teeth in the driven gear is not divisible by the number of teeth in the driving gear.

Spindles, Shafts, Plungers and Rods

A spindle is usually made of steel, and very often of a higher grade of steel, since in most cases it supports tools, cutters or work pieces. High accuracy and, in most cases, high strength are required. The majority of spindles require only localized hardening to protect against wear and chipped or marred surfaces.

There are cases where the shape of the spindle is such that the whole spindle should be hardened, but even a long, slender spindle could be locally hardened because it is then easy to straighten the spindle, if necessary, before grinding.

Two good steels are:

a. AISI 52100, a high chrome carbon steel quenched in oil and drawn to a hardness of 60 R_c. It has a high resistance to wear.
b. AISI 6145, a chrome vanadium steel quenched in oil to a hardness of 50 to 55 R_c. It has a high resistance to fatigue.
This steel may also be used for large spindles and may be surface hardened or induction hardened to 65 to 75 scleroscope with a minimum of distortion. It has good resistance to wear.

Steel shafts used just for transmission of motion from one part of the machine to another part usually require no hardening. In fact, in most cases it would be just a waste of time and money. Consider each case thoroughly and provide for heat treatment only if there is a question of wear, or if you are limited to size, and the required strength can only be obtained by heat treatment.

If balls or rollers are in direct contact with the shaft, if cams or other machine components are frequently replaced, or if the shaft has splines or integral keys in sliding contact under load, localized hardening is required. (Figure 4.7–1)

The shape, length, diameter, speed and load may also be such that heat treatment should be applied to the entire shaft because there is a considerable increase in strength with heat treatment of alloy steel. (Figure 4.7–2)

Rods and plungers are also, as a rule, made of steel, and the same heat treatment consideration can be applied as to shafts. Plungers in machine tools are generally small, and heat treatment of the entire

4.7 Material Selection and Heat Treatment

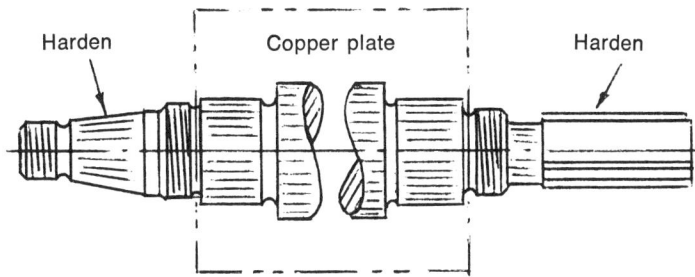

Figure 4.7-1. Selective heat treatment of alloy steel shaft.

Table 4.7-2. Ultimate Tensile Strength of Steel

Brinell 3000 kg to Ultimate Tensile Strength for Steel				
Brinell 3000 kg Load 10 mm Ball	Ultimate Tensile Strength, psi		Brinell 3000 kg Load 10 mm Ball	Ultimate Tensile Strength, psi
200	100,000		400	215,000
225	108,000		425	227,000
250	122,000		450	238,000
275	141,000		475	249,000
300	158,000		500	258,000
325	174,000		525	267,000
350	188,000		550	282,000
375	202,000		575	295,000
			600	308,000

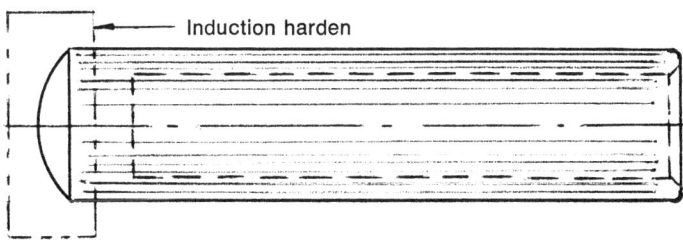

Figure 4.7-3. Induction hardening of plunger

Prototype Machine

part often is the best. If the plunger is long and slender and only the end needs to be hardened, localized heat treatment may be applied. (Figure 4.7–3)

Generally speaking, shafts and rods running and sliding at moderate speeds and loads, and in contact with dissimilar materials such as cast iron or copper alloys, need not be hardened, provided lubrication is adequate.

CASE HARDENED ALLOY STEEL. This steel may be selectively hardened by:

a. Copper plating the portion where no hardening is desired, before the carburizing process. This will keep the carbon from entering this portion of the steel, and therefore no hardening takes place when the part is quenched after being heated to the proper temperature. (Figure 4.7–1) It is also then possible to straighten the shaft before grinding the hardened surfaces.

b. As shown in Figure 4.7–4, a false section may be prepared to the portion where no hardening is desired, before the carburizing process. This false section may then be removed after the carburizing process. The carbon penetrates the steel only to a depth of a few thousandths of an inch, depending on the length of time in the carburizing furnace, so a false section of a maximum of ⅛ inch should be adequate. When the steel has reached room temperature after the carburizing process, the false section can easily be removed with standard tools, and after the steel has been reheated and quenched only the portion that has been carburized will reach maximum hardness of about 60 R_c.

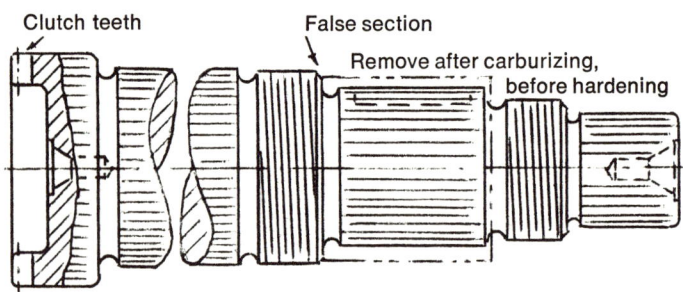

Figure 4.7–4. Selective heat treatment of alloy steel shaft by adding a false section

4.7 Material Selection and Heat Treatment

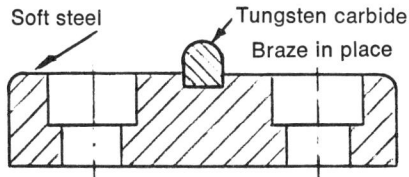

Figure 4.7-5. Tungsten carbide cam rider

When shock loads are encountered, case hardened steel is a good selection, because it possesses a ductile and tough core and has a high resistance to impact, with a high surface hardness.

OIL HARDENING STEELS. This type of steel may be selectively hardened with a torch or induction coil and is a good selection for shafts, rods and plungers when a hardness of 45 to 55 R_c is sufficient.

Levers, Connecting Links and Minor Parts

Levers, links and covers are often made of cast iron because their appearance is good and the finished portions can be processed at low cost. If appearance is not important, low carbon steel may be used, which can easily be flame cut to the desired shape. Minor parts, such as washers, studs, bolts, nuts, collars, spacers and pulleys, are usually made from free cutting steel and left in the soft state. If hardening is required, as for a washer used under a screw which must be frequently loosened and tightened again, oil quenching and drawing to a hardness of 30 to 40 R_c is usually adequate.

For minor parts subject to hammer blows, such as dogs, pawls, trip pins, latches, ratchet wheels and clutches, a case hardened nickel alloy steel like AISI 2315, carburized not deeper than ⅓ of the total sectional thickness, and hardened and drawn to a hardness of 55 to 58 R_c, should be used. This steel can also be hardened to a maximum hardness of 60^+R_c, which may be suitable for cam rollers. When a cam rider is required, high speed steel may be used which can be hardened to 61 to 65 R_c. A tungsten carbide rider brazed to a soft steel body may also be used. (Figure 4.7-5)

The hardest material should always be on the small area, when a large area passes over a small area or a small area passes over a large area. Well lubricated contact surfaces are always desirable.

Moving Members in Contact where No Lubrication Is Available

Carbon materials or graphite suspended materials may be used for

Table 4.7–6. Common Materials Used in Machine Tools

Material	Used For	Hardness as Purchased	Hardness in Use
Grey cast iron	Frames, housings, brackets, levers, caps, covers, cams		May be chilled for acute cams
Nodular cast iron	Part requiring high strength Good for impact loads		Up to 55 R_c
Cast aluminum	Guards, doors or parts frequently handled		
Aluminum tubing	Lubricating, hydraulic and coolant systems		
Dead soft aluminum	Hydraulic gaskets		
SAE #62 bronze	Bushings and guide parts		
SAE #65 phos. bronze	High speed, heavy duty worm gears, special gears		
SAE #68 cast al. brze.	Heavily loaded worm gears, wear plates		
AISI #2315 nickel steel	Parts subject to hammer blows	163 to 192 Brinell	55 to 58 R_c Rc
AISI #2315 nickel steel	Parts where maximum hardness is required	163 to 192 Brinell	60+ R_c
AISI #8617 Cr Ni Mo stl.	Gears and similar parts	143 to 187 Brinell	60+ R_c
AISI #C1141 steel	Shafts, adjusting screws, plungers, small parts	179 to 228 Brinell	45 to 55 R_c
AISI #C1141 steel	Bolts, nuts, shoulder screws, studs, etc.	179 to 228 Brinell	30 to 40 R_c
AISI #4140 Cr Mo steel	Clamp nuts, and screws and adjusting screws	179 Brinell	38 to 42 R_c
AISI #4140 Cr Mo steel	Clutches, studs, stop screws, plungers, stop plates	179 Brinell	50 to 55 R_c
AISI #4640 Mo steel	Gears and similar machine parts	Good machinability	60 to 75 scleroscope

4.8 Transmission of Motion

Material	Used For	Hardness as Purchased	Hardness in Use
AISI #52100 High Cr C stl.	For spindles and parts of high strength and hardness	Fair machinability	60 R_c
AISI #6145 Cr V steel	For spindles of high fluctuations in load	Fair machinability	50 to 55 R_c
Wrought nitriding stl.	Fine pitch gears, heavily loaded parts, good wear	163 to 192 Brinell	
AISI B1112 steel	Screws, nuts, collars and spacers	Good machinability	Soft
AISI #1020 steel	Screws, nuts, collars and parts for fabrication	Good machinability	Soft
Tool steel	Dies, tools and small machine parts for good wear		60 to 62 R_c
High speed steel	Cutting tools, small centers, cams and riders		61 to 65 R_c
Tungsten carbide	Cam riders		92 R_a

contact surfaces where speeds are reasonable and only atmospheric moisture or liquids other than lubricating oils are available. These materials have excellent hardness and will give good results under normal operating conditions.

Many plastic materials are now available which will give good results with little or no lubrication, provided that the speeds and loads are moderate. These materials are good where contamination of parts being processed cannot be tolerated.

Some of the most common materials used in machine tools are shown in Table 4.7–6.

4.8 Transmission of Motion in Machine Tool Design

In machine design, many conditions govern the transmission of a force from one part of a machine to another. Given here are some examples of fundamental rules which are supported by actual examples of successful design.

Table 4.7–7. Rockwell$_c$ to Brinell Conversion Table

Rockwell$_c$ 150 kg to Brinell 3000 kg
For Hardened Steel and Alloys

Rockwell$_c$ 150 kg load Brale penetrator	Brinell 3000 kg load 10 mm Ball	Rockwell$_c$ 150 kg load Brale penetrator	Brinell 3000 kg load 10 mm Ball
60	614	45	426
59	600	44	415
58	587	42	393
57	573	40	372
56	560	38	352
55	547	36	332
54	534	34	313
53	522	32	297
52	509	30	283
51	496	28	270
50	484	26	260
49	472	24	250
48	460	22	240
47	448	20	230
46	437		

The straight line in geometry is the shortest distance between two points. Similarly, in machine design, the most direct way is the best in transmitting motion from one part of the machine to another. This applies to rotating drive as well as reciprocating motion. Therefore, in working out a drive diagram for an efficient, economical and accurate motion, you should strive for as few members as possible and the shortest possible distance.

Machine Slides

A slide is one of the main components for transmitting force to the point of application. In moving a heavy sliding member, it is better to apply the force directly through a cam roller (Figure 4.8–1) than through a lever arrangement (Figure 4.8–2) unless you have some definite reason not to do so; for instance, to provide for an adjustable stroke where a lever or some other such arrangement would have to be

4.8 Transmission of Motion

Table 4.7–8. Rockwell$_b$ to Brinell Conversion Table

Rockwell$_b$ 100 kg to Brinell 500 and 3000 kg For Unhardened Steel of Soft Temper, Grey and Malleable Cast Iron and Nonferrous Metals

Rockwell$_b$ 100 kg ld. 1/16 inch diam. ball	Brinell 500 kg ld. 10 mm ball	Brinell 3000 kg. ld. 10 mm ball	Rockwell$_b$ 100 kg ld. 1/16 inch diam. ball	Brinell 500 kg ld. 10 mm ball	Brinell 3000 kg. ld. 10 mm ball
100	201	240	66	104	117
99	195	234	64	101	114
98	189	228	62	98	110
97	184	222	60	95	107
96	179	216	58	92	104
95	175	210	56	90	101
94	171	205	54	87	
93	167	200	52	85	
92	163	195	50	83	
91	160	190	48	81	
90	157	185	46	79	
89	154	180	44	78	
88	151	176	42	76	
87	148	172	40	74	
86	145	169	38	73	
85	142	165	36	71	
84	140	162	34	70	
83	137	159	32	68	
82	135	156	30	67	
81	133	153	28	66	
80	130	150	24	64	
79	128	147	20	62	
78	126	144	16	60	
77	124	141	12	58	
76	122	139	8	56	
75	120	137	4	55	
74	118	135	0	53	
72	114	130			
70	110	125			
68	107	121			

Prototype Machine

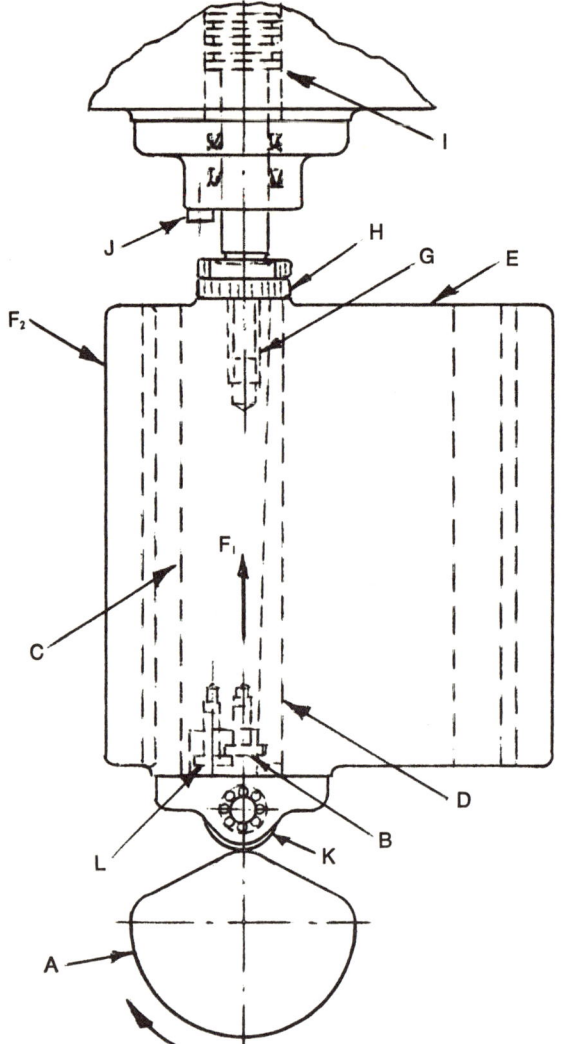

Figure 4.8–1. Machine slide, cam and roller operated.

A cam A is transmitting a force F_1 to the slide E, through roller K. The roller is kept in contact with the cam by a hydraulic load applied to piston I to insure the accuracy of the cam being transmitted by the slide without fluctuations. A long and narrow guide C insures directional stability. The sliding clearance of the guide is adjusted with tapered gib D through screw B, and locked with screw L. If slide must be held at end of the stroke to great accuracy, it may move to stop J. Collar H can be ground to suit the requirements and locked in place with threads G on piston rod. If the slide must be held against stop for a long time, this is best done with hydraulic power. F_2 represents force from tools or parts being processed.

4.8 Transmission of Motion

used. But just to add the lever so you can find a more convenient place for the cam is not good practice. Arrangements would have to be made in the design so that the cam could easily be changed for repair or replacement.

In Figure 4.8–1, the force transmitted from the cam to the sliding member is applied through a roller fastened to the sliding member. This is good from the standpoint of lowering frictional resistance; but in many cases where very high accuracy is required, the roller will introduce inaccuracy. Whether you use anti-friction or plain bearings,

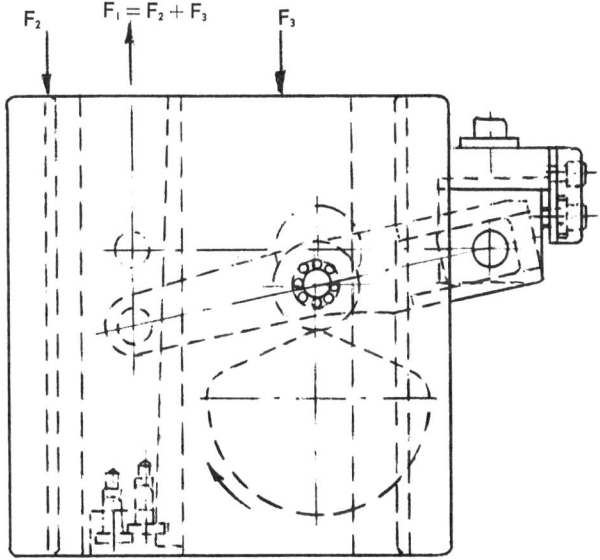

Figure 4.8–2. Machine slide, cam and lever operated.

Good control is always important. The long, narrow guide in the slide, gives good directional control for the slide. The end of the lever operating the slide is fastened to the center of the guide by a pin. Only rotational motion occurs here, giving good control for the lever.

The stroke of the slide is adjustable at the fulcrum pin which is located in a rectangular sliding block. The rectangular sliding block thus slides in the lever when stroke adjustment is made. Because the other end of the lever must travel in a straight line when moved by the cam, a small amount of sliding occurs between the sliding block and the lever at the fulcrum. It is, therefore, best to have the cam turn toward the fulcrum in this case. In cases where only rotational motion occurs at the fulcrum, the cam should turn away from the fulcrum for best results.

Forces F_2 and F_3 are counter balanced at the center of the guide by force F_1. See figure 4.8–7.

Prototype Machine

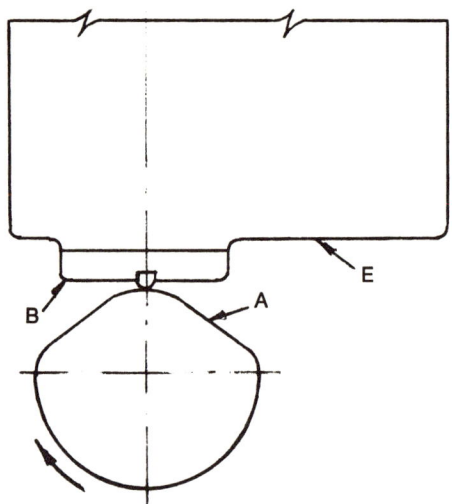

Figure 4.8-3. Machine slide, cam and rider operated.
To insure accuracy in repetition, a cam rider may be used. Cam A moves slide E through a tungsten carbide rider fastened in block B.

you will have inaccuracy. You are expecting perfect repetition when you come to a certain point on the cam, but you find that the repetition is not as close as desired because the cam does not transmit force through the same diametral points on the roller every time. Much time and money can be wasted trying to correct this trouble by perfecting the roller and its mounting. Again, this is a case where the most direct contact is best for high accuracy.

A cam rider is much better when high accuracy is required. (Figure 4.8-3) Accuracy is designed and manufactured into the cam. The relationship of hardness of cam rider to cam surface should be such that, considering speed, forces applied and other factors, the cam rider will wear first. In most cases this will call for a much harder cam rider because, as a rule, the forces are applied in sequence from a much larger area of the cam.

The cam rider should be kept in close contact with the cam at all times so the accurate motion of the cam can be transmitted without bounce. This may be done with springs, hydraulic power or pneumatic power, depending on circumstances.

If a sliding member is advanced to a predetermined position by a

4.8 Transmission of Motion

cam and repetitive accuracy is important, the simplest way to provide it is to let the sliding member come to an adjustable stop. (Figure 4.8-1) By putting the slide material under compression, it will act as a spring and insure repetition. It is important to put the stop in the center of the guide or as close to this point as possible so the slide will be under compression, or close to it. If the motion of the slide is provided for by hydraulic or pneumatic power instead of by a mechanical cam, you do, of course, need a mechanical stop to assure accurate repetition.

Other factors also control accuracy and performance when transmitting motion in machine tools. The sliding member must be under good control. This is accomplished by placing the guide of the sliding member as close to the applied and transmitted forces as practical. (Figure 4.8-1) This is especially important when you have a plain bearing guide.

If you have an anti-friction guide, you can stay farther away from the guide with the applied and transmitted forces, and still have good accuracy, performance and control.

A plain guide must be long in proportion to the width. A good rule of thumb is a ratio of between 4 and 7 to 1. Anti-friction guides can be much shorter, depending on the forces transmitted. The applied force can be a considerable distance from the guide if the guide is rolling on preloaded steel balls.

The selection of a guide for a machine slide requires careful consideration, and is influenced by numerous conditions. The ideal design would appear to be a plain cast iron narrow guide, well lubricated, located at the line of action. Where this is not possible, anti-friction bearings may be applied to the guide. There are many variations of these bearings, some of which are shown in Figures 4.2–18 and 4.2–25. Using such bearings may be perfectly all right where loads are light, or when the accuracy of the parts being processed is not affected. But in some cases where the forces generated by the tools in contact with the part being processed are high and of an intermittent nature, the anti-friction balls or rollers may act as bridges, generating chatter marks on the part being processed. In such a case, a compromise may have to be worked out if it is impossible to have the operating force applied at the most advantageous point.

Because there is not always a clearly defined use of the words

Prototype Machine

force and power in industry, it is well to review here Newton's three fundamental laws of motion:

1. A body will remain in a state of rest or continue at a uniform rate of motion in a straight line until a force acts on the body.
2. A force acting on a body produces an accelerated motion of the body in the direction of the force, and the magnitude of the acceleration is directly proportional to the force, and inversely proportional to the mass of the body.
3. To every action there is an equal and opposite reaction.

Force, therefore, can be defined as the incentive that produces or leads to a change in motion. A single force or a system of forces may be required to keep a body in equilibrium.

The mass of a body is expressed as the weight of the body in relation to the gravitational force, $M = \dfrac{W}{g} = \dfrac{W}{32.17}$, and the force can then be expressed as, $F = M \times a$ where a is the acceleration in feet per second per second.

The velocity of a body is expressed as distance moved in feet (s) in relation to the time required in seconds (t), $v = \dfrac{s}{t}$

Acceleration then, according to Newton's second law of motion, is proportional to the resultant of the forces that cause the acceleration:

$$a = \frac{F}{M} \text{ or } \frac{v}{t}$$

A body may thus be accelerated from rest at a uniform rate of feet per second in time, t, $v_o = a \times t$; or after having attained v_o velocity, an additional force may increase the attained velocity v_o uniformly to a new velocity $v = v_o + a \times t$ (Figure 4.8–4)

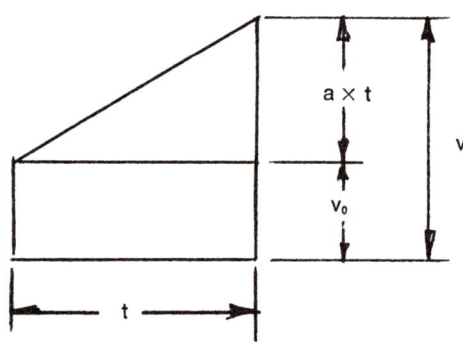

Figure 4.8–4.

4.8 Transmission of Motion

The distance traveled, s, may then be found by solving for the area of the surface: $s = \dfrac{v_o + v}{2} \times t$

Acceleration is positive for increasing velocity. Deceleration is negative for decreasing velocity. The displacement of a body in a rectilinear motion relative to a reference point, the velocity of a body, which is the time rate of displacement relative to this reference point, and the acceleration of a body expressed in time rate of velocity change are all vector quantities because they all have direction and magnitude.

The force causing displacement, velocity and acceleration of a body from rest to v_o velocity or from v_o to v velocity may be the result of several forces. This resultant force R is then determined using the fundamental principle of triangle laws, and may be expressed by vectors in a parallelogram. (Figures 4.8–5a and 5b)

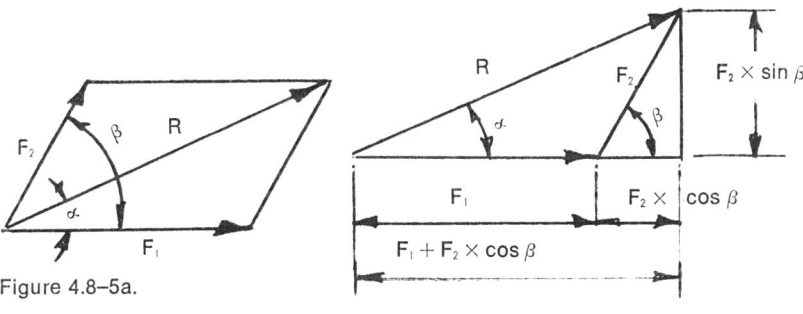

Figure 4.8–5a.

Figure 4.8–5b.

F_1 and F_2 are concurrent forces working in the same general direction as the resultant R. Vector R shows graphically the direction and magnitude of the resultant force and may be determined graphically by measuring the vectors or by calculations. Referring to Figure 4.8–5b, which is the solution of force parallelogram, and dividing Figure 4.8–5a into two triangles, we have:

$$R = \sqrt{(F_1 + F_2 \times \cos \beta)^2 + (F_2 \times \sin \beta)^2}$$
$$= \sqrt{F^2_1 + F^2_2 + 2\, F_1\, F_2\, \cos \beta}$$

Forces F_1 and F_2 in figure 4.8–5a are concurrent forces. Most practical applications for machine tools have concurrent and nonconcurrent forces as with a cam or screw, moving a slide by force F_1 and being opposed by the force of the tools against the work piece F_2. (See figure

Prototype Machine

4.8-1) The graphical layout or algebraic method of solution is just as simple in solving for the resultant vector R. (Figure 4.8-6)

$$R = \sqrt{F^2_1 + F^2_2 - 2 \times F_1 \times F_2 \times \cos \alpha}$$

A machine slide may also be moved by a force F_1 to overcome two parallel, opposing forces F_2 and F_3. (Figure 4.8-2) It is assumed that forces F_1, F_2 and F_3 are in the same plane. By making graphic vector diagrams of F_2 and F_3, the resultant vectors R_1 and R_2 may be found, showing where the minimum force F_1 should be applied to keep the slide in equilibrium. As in algebra, adding a plus and a minus value of the same magnitude, will not alter the result. Therefore, as shown in Figure 4.8-7, two forces F_4 in exactly opposite directions, will not impart any change to forces F_2 and F_3.

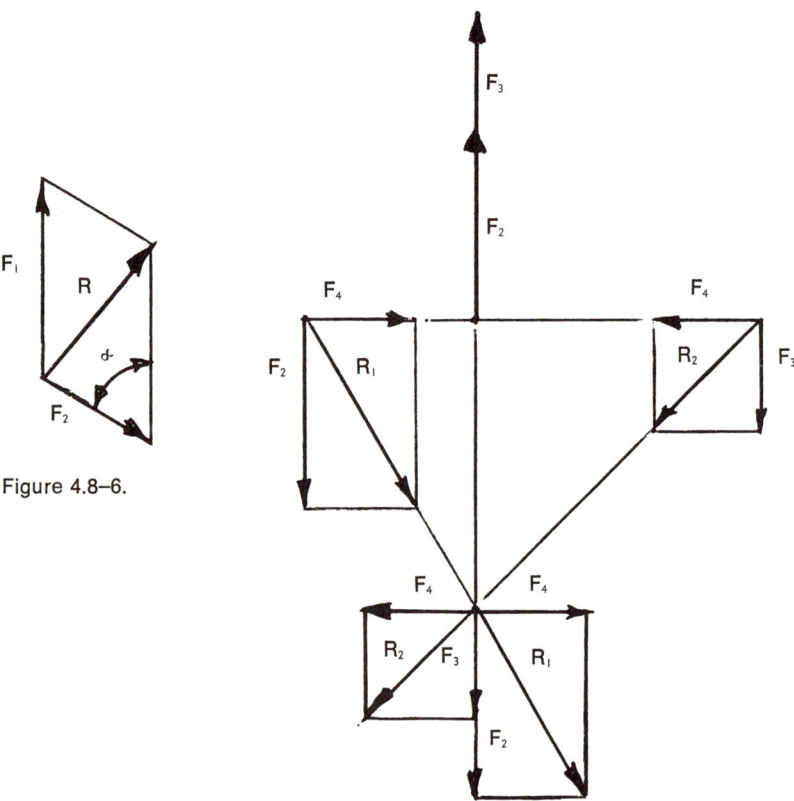

Figure 4.8-6.

Figure 4.8-7.

4.8 Transmission of Motion

F_2 and F_3 may, therefore, be transformed to two concurrent forces R_1 and R_2 respectively, and $F_1 = F_2 + F_3$. If applied at the intersection of R_1 and R_2, the point of action, the forces will be kept in equilibrium. This is the ideal location for the center of the guide in the slide.

The slide is kept in directional alignment with the guide. The most common way of adjusting the sliding clearance is with a tapered gib. The guide then gives the slide directional stability. Working stability is obtained by two surfaces placed at a distance to satisfy stability for the load conditions. If the guide is some distance away from the edges of the slide, the surfaces are placed on either side of the guide, and if operating planes or load conditions make it necessary, the slide is kept in place by flat gibs. (Figure 4.8–8) If the guide is at one side of the slide, however, one side of the guide may be for working stability and the other two for directional stability. (Figure 4.8–9)

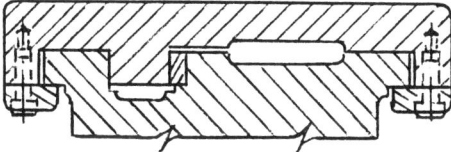

Figure 4.8–8. Sectional view of machine slide

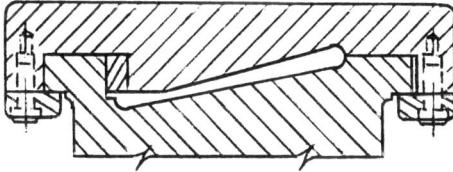

Figure 4.8–9. Sectional view of machine slide

Wedges

A wedge may be a fine and simple way of adjusting one machine member in relation to another for short distances, but when you are moving under power, transmitting a force from one member to another, a lever arrangement is much better. If you are considering the use of wedges for motion under power, make sure you have investigated the efficiency of the wedge arrangement as compared to levers.

Levers

There are so many different arrangements of levers that no attempt

Prototype Machine

will be made here to describe some of the ingenious arrangements, because each case, in the course of designing a machine, would call for an arrangement peculiar to itself.

Every new machine seems to require mechanisms that are a little different from the mechanisms you have used before. In investigating applications of unusual mechanisms your search will at least serve to stimulate your thinking if it does not actually turn up a new method of application.

It is important to consider the necessary fundamental requirements in lever design. Excessive sliding under contact pressure in a lever should be avoided if possible. Careful study of sliding under pressure will reveal ways to minimize or almost eliminate it.

The upper part of Figure 4.8–10a shows a lever operating a plunger. The contact surface of the lever is radial to the center of the lever, which is good. The stroke of A of the plunger is, however, some distance above the center of the lever. This arrangement produces a sliding contact C.

The lower part of the figure shows stroke B to be of same length as

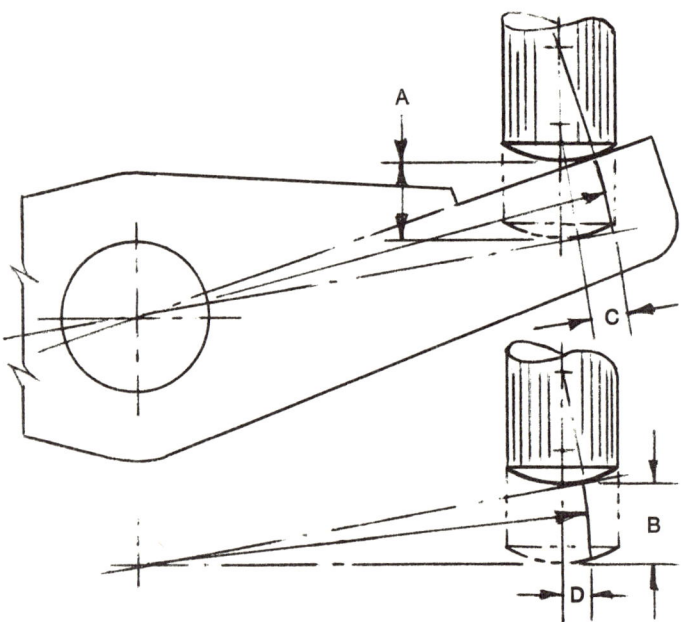

Figure 4.8–10a. Lever operated plunger sliding under contact

4.8 Transmission of Motion

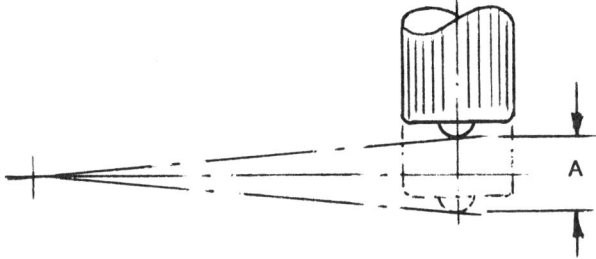

Figure 4.8–10b. Lever operated plunger, practically no sliding under contact

stroke A, originating at the center line. This has improved the sliding very little as shown at D.

As shown in Figure 4.8–10b, the design has been improved considerably. The stroke A is here divided equally above and below the center line. The plunger has also been provided with a hardened button of a small contact radius. These changes result in reduced sliding, so small that it has no practical significance.

A more direct approach than a lever for transmitting motion of minor importance in a rectilinear or angular fashion is by using balls confined in a tube, figure 4.8–11.

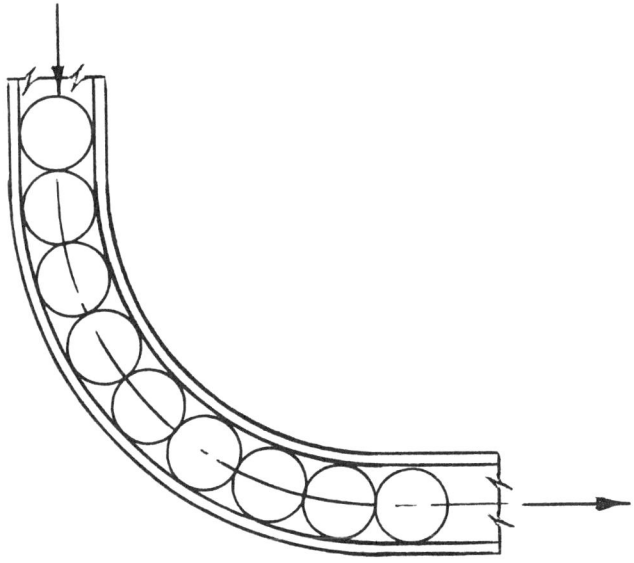

Figure 4.8–11. Transmitting motion with balls

Prototype Machine

Screws

Ball bearing screws have attained some use in machine tool design especially for light power drives. Life may now be fairly accurately predicted, but close consideration should be given to the application. (Figure 4.2–19)

Screws used as translating screws, either for adjustments or power transmission, have American standard threads with 30° pressure angle or Acme threads with 14½° pressure angle.

The following observations should be considered:

1. When the plane of rotation of the control member is perpendicular to the plane of motion of the controlled member, the motion of the controlled member should be away from the control member when the rotation is clockwise, whether the plane of motion is horizontal, vertical or angular. (Figure 4.8–12)

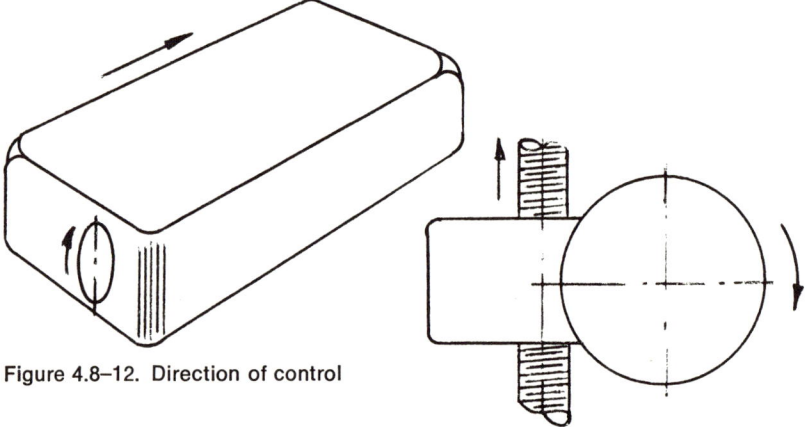

Figure 4.8–12. Direction of control

Figure 4.8–13. Direction of control

2. When the plane of rotation of the control member is parallel to the plane of motion of the controlled member, the motion of the controlled member should be as shown in Figure 4.8–13 for clockwise rotation of the control member, whether the plane of the controlled member is horizontal, vertical or angular.

The efficiency of a screw is very low for a helix angle under 20°. The best efficiency is between 30° and 60° helix angle, and the efficiency drops rapidly, for a helix angle above 60°. It is therefore good

4.8 Transmission of Motion

practice to keep the diameter of the screw as small as other considerations will permit. This has been proved by actual experiments.

The following formula may be used for approximate results for a screw having helix angle between 30° and 60°.

To lift a weight vertically or move a weight horizontally, when W is the total weight or opposing force to be moved, r is the pitch radius of the screw, α is the helix angle, β is the included pressure angle, R is the radius of the lever or hand wheel and μ is the coefficient of friction:

$$F = \frac{W \times r}{R} \left(\frac{\mu \times \cos \alpha}{\cos \frac{\beta}{2}} + \tan \alpha \right), \text{ all dimensions in pounds and inches.}$$

To lower a weight vertically, the formula would be:

$$F = \frac{W \times r}{R} \left(\frac{\mu \times \cos \alpha}{\cos \frac{\beta}{2}} - \tan \alpha \right)$$

Shafts

When rotary motion of the control member is translated to rotary motion of the controlled member, the rotation of the controlled mem-

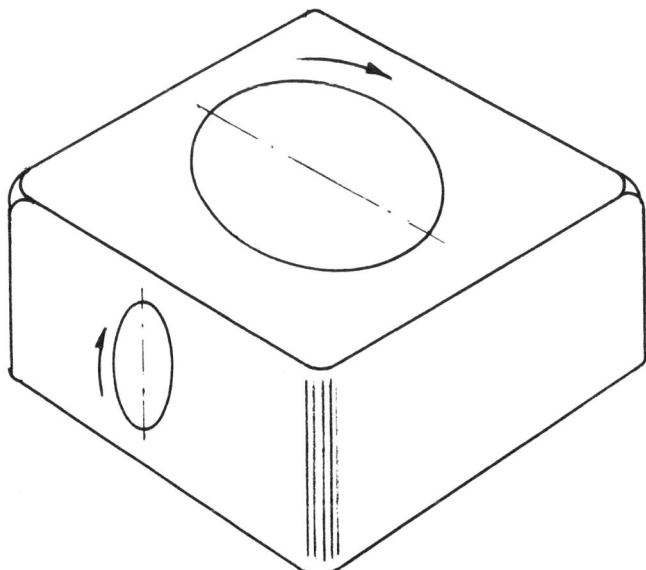

Figure 4.8–14. Direction of control

Prototype Machine

ber should be in the same direction as the control member as viewed. (Figure 4.8–14)

In the first part of this chapter we considered force as applied to machine members. Considering members in motion, however, we must consider the work being performed, which is the product of force and distance and in most cases for machine design would be measured as inch pounds or foot pounds.

Power then is the work performed in a certain length of time or the product of force and distance divided by time, expressed in foot pounds per minute or foot pounds per second.

Horsepower is the unit of power for engineering work. One horsepower is 33,000 foot pounds per minute or 550 foot pounds per second.

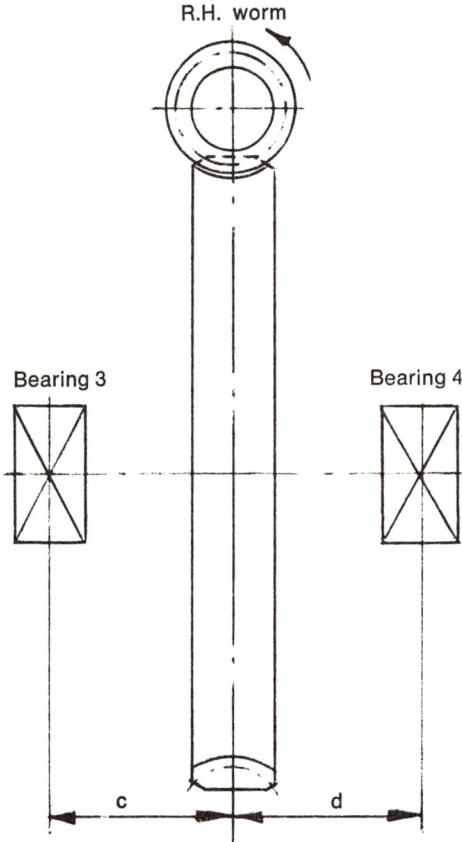

Figure 4.8–15a. Worm and worm gear bearing calculations

4.8 Transmission of Motion

When figuring power transmitted by a shaft, the unit torque, T, is used.

$$T = \frac{63{,}025 \times \text{hp}}{N}, \text{ where N is the rpm of the shaft.}$$

In figuring the loads that the gears and bearings are subject to on a shaft, the horsepower of the main drive motor must be used and the average life factor should be approximately 25,000 hrs. Since a worm gear drive presents some problems similar to those of other types of spiral gear application when figuring bearing loads, we will figure the bearing loads referring to Figures 4.8–15a, 4.8–15b and 4.8–15c.

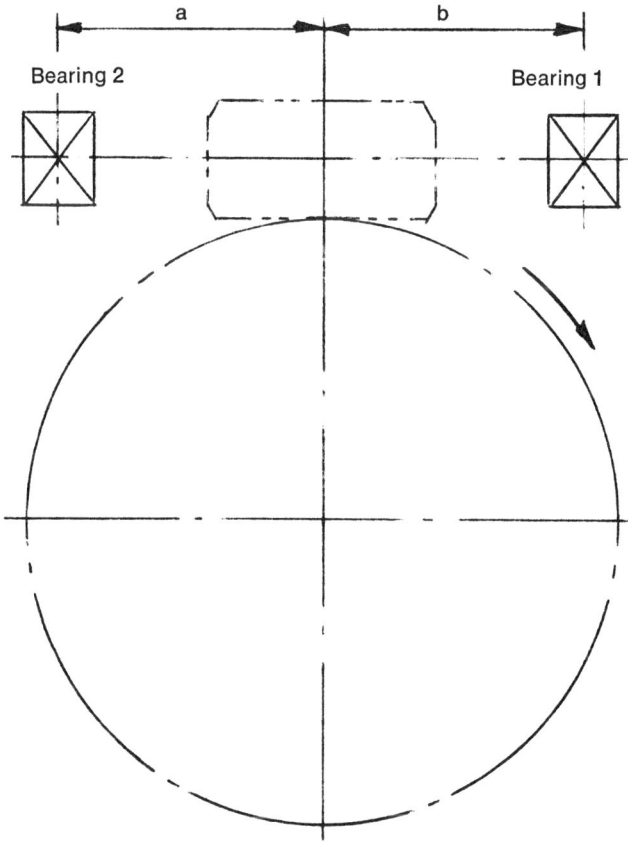

Figure 4.8–15b. Worm and worm gear bearing calculations

Prototype Machine

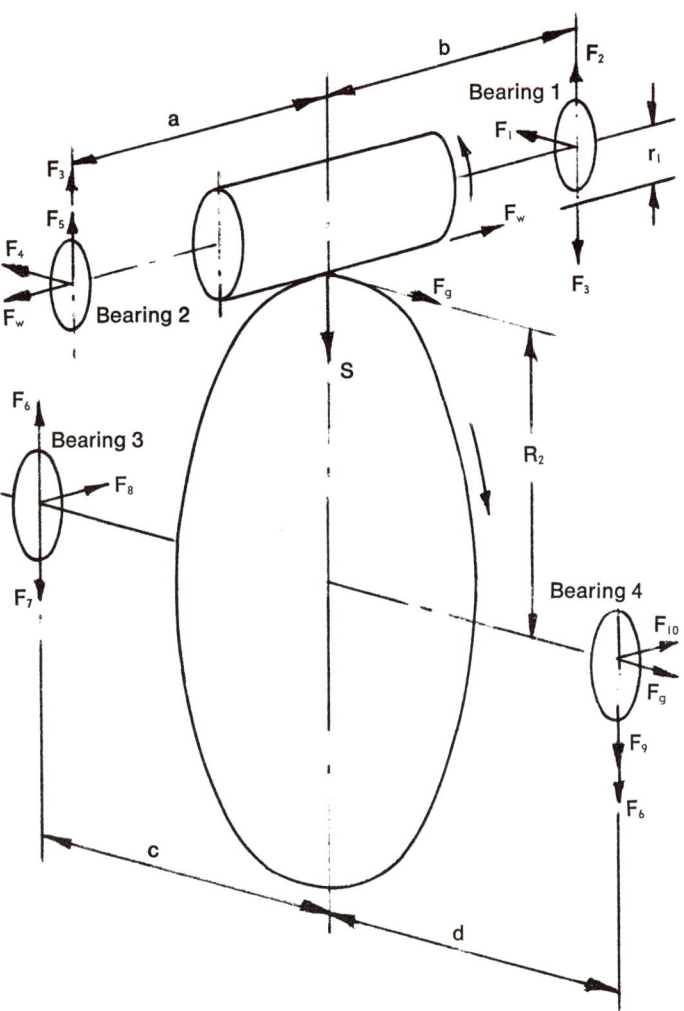

Figure 4.8–15c. Worm and worm gear load vectors.

Because only one bearing on the worm shaft can take the thrust load F_w, and only one bearing on the worm gear shaft can take the thrust load F_g, the other bearing on both shafts can be a radial bearing, if this will simplify the design.

To simplify the presentation, only the pitch cylinder of the worm has been shown. The pitch circle of the worm gear is shown as an ellipse. All the bearings are shown as ellipses at the center of each bearing.

4.8 Transmission of Motion

$$F_g = \frac{T}{r_1} = \text{tangential force of worm}$$

$r_1 =$ Pitch radius of worm, inches
$r_2 =$ Pitch radius of worm wheel, inches

$$S = \text{Separating force} = \frac{F_1 \times \tan \alpha}{\tan \gamma}$$

$\alpha =$ Worm tooth pressure angle
$\gamma =$ Lead angle of worm

In analyzing the thrust loads on the worm and worm gear shafts derived from the rotation specified, it will be seen that the thrust on the worm shaft, F_w is in the direction of bearing 2. The thrust on the worm gear shaft, F_g is in the direction of bearing 4. In both cases only one bearing on the shaft can take the thrust load.

$$F_w = \frac{F_g}{\tan \gamma} = \text{Worm thrust}$$

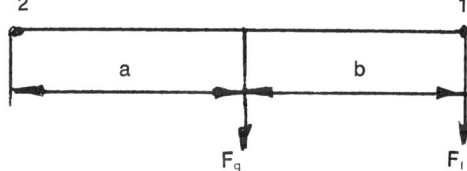

$$F_1 = \frac{F_g \times a}{a + b}$$

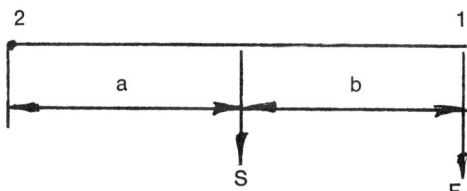

$$F_2 = \frac{S \times a}{a + b}$$

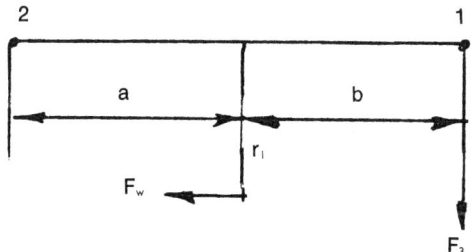

$$F_3 = \frac{F_w \times r_1}{a + b}$$

Total radial load on bearing $1 = \sqrt{F^2_1 + (F_2 - F_3)^2}$

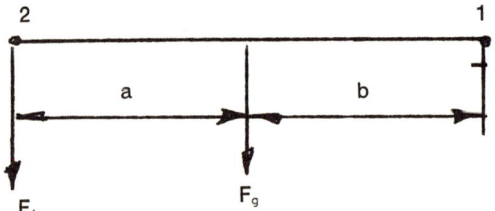

$$F_4 = \frac{F_g \times b}{a + b}$$

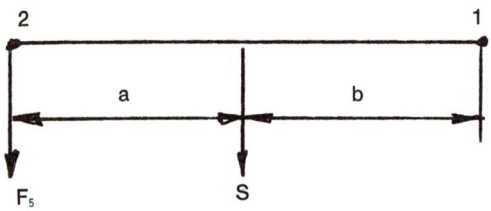

$$F_5 = \frac{S \times b}{a + b}$$

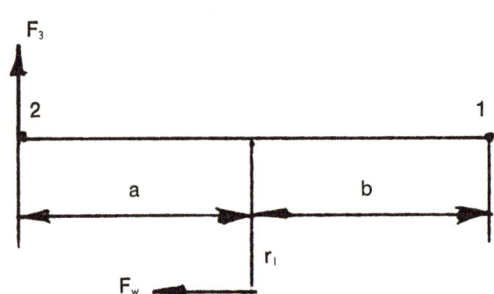

$$F_3 = \frac{F_w \times r_1}{a + b}$$

$$R_2 = \sqrt{F^2_4 + (F_5 + F_3)^2}$$

Total radial load on bearing $2 = \sqrt{R^2_2 + F^2_w}$

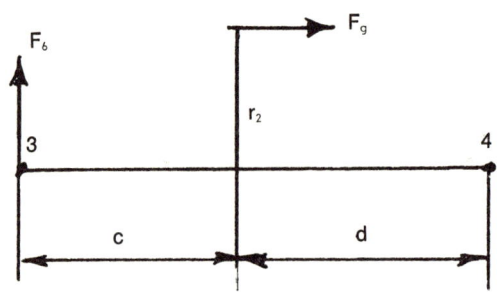

$$F_6 = \frac{F_g \times r_2}{c + d}$$

4.8 Transmission of Motion

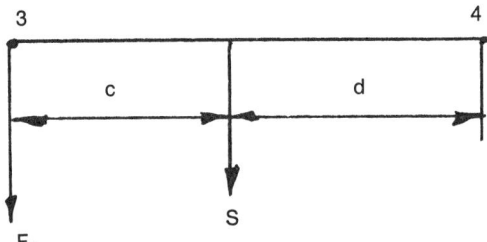

$$F_7 = \frac{S \times d}{c + d}$$

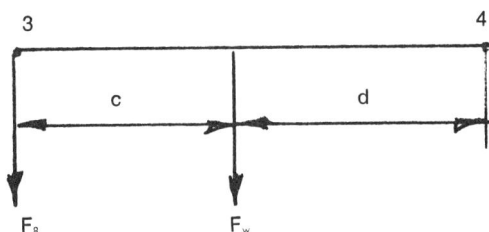

$$F_8 = \frac{F_w \times d}{c + d}$$

Total radial load on bearing $3 = \sqrt{F^2_8 + (F_6 - F_7)^2}$

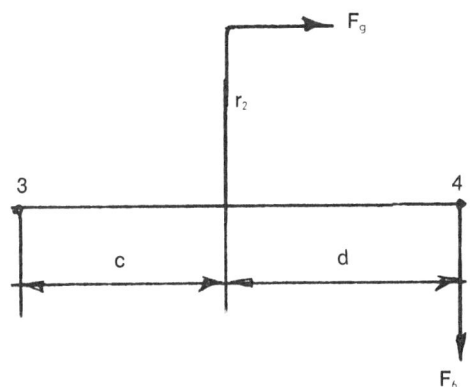

$$F_6 = \frac{F_g \times r_2}{c + d}$$

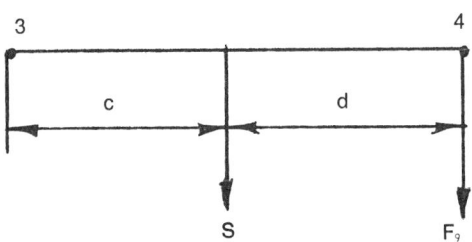

$$F_9 = \frac{S \times c}{c + d}$$

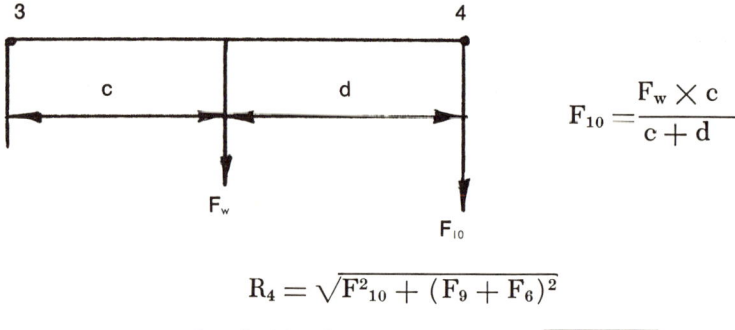

$$R_4 = \sqrt{F^2_{10} + (F_9 + F_6)^2}$$

Total radial load on bearing $4 = \sqrt{R^2_4 + F^2_g}$

Figure 4.8–16 shows direction of thrust for spiral gears on parallel shafts. Reversing the rotation of the gears, reverses the thrust.

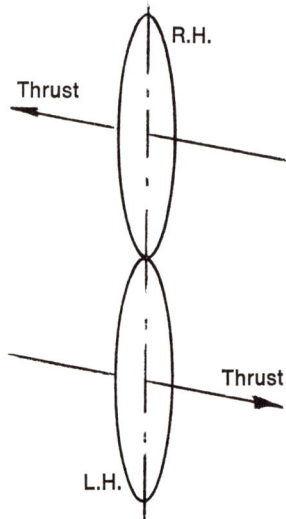

Figure 4.8–16. Direction of thrust for spiral gears

4.9 Machining Practices for Machine Tool Design

It is of utmost importance for a machine tool designer to be familiar with machining practices. In every step of your planning, you should have a good idea of where and how each part can be made. Although this book cannot go into all methods of manufacturing, it takes up a few things which will be helpful in your planning.

4.9 Machining Practices

Milling machines are used extensively for finishing parts having plain surfaces. Surface flatness, angularity and smoothness can be controlled very well in a milling machine and held to accuracy close enough for most work. It is also considered an economical method of finishing.

Planers and shapers are used where a single tool is required.

Boring mills are used for finishing large parts having circular shapes such as holes and pilots.

Parts having external and internal diameters are, of course, finished by turning in a lathe, or if greater accuracy is required, by grinding.

Many small parts, and especially parts that are thin in proportion to area, are preferably finished by grinding. Where great accuracy and smoothness are desired, and repetition of accuracy from part to part is required, this is an excellent method. Iron and steel parts can then be held down magnetically.

Many very excellent machines of different types are now available for finishing holes to great accuracy, and to very close tolerances measured from hole to hole.

Drill presses or radial drills are used for holes in general. When specifying drilled holes, always remember that long holes of small diameter are difficult to drill and may have considerable runout. A drilled hole where the drill is unsupported on one side for part of the hole also causes breakage of the drill.

Always, in your design planning, keep in mind that if a machine part can be completely finished in one machine and in one clamping this is by far the most economical. This cannot always be done, but should be your aim. Greater accuracy is possible with few changes in setups. Many times with just a few minor changes in your design, you can improve finishing methods without sacrificing anything in your design.

An interrupted cut in turning is bad practice and should be avoided. There is usually some way in which your design can be changed to avoid this condition, and quite often you improve the design by doing so. Just one simple example will stimulate your thinking about changes in more complicated cases. (Figure 4.9–1) As originally planned, this part was made of cast iron in one piece. The hole B and pilot A were required to be concentric and square with surface E. The best arrangement, therefore, was to bore the hole B, turn the pilot A

Prototype Machine

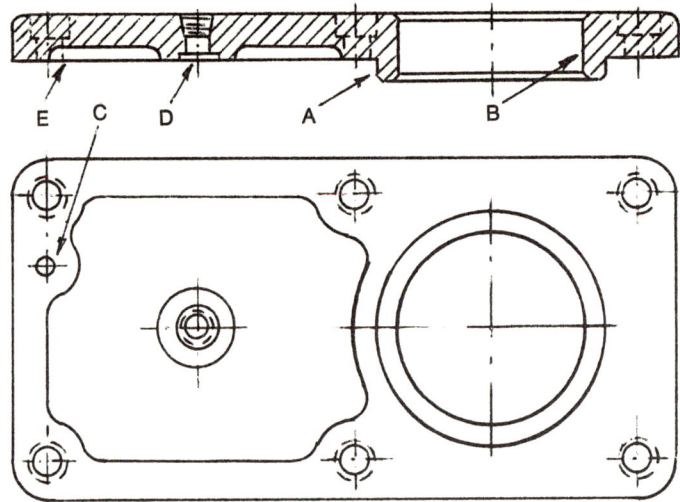

Figure 4.9-1. This cast iron part has been designed with a pilot A to make sure that hole B is centered with hole in mating bracket. Dowel C insures that hole D is properly lined up with part in mating bracket.

This is all right for proper assembly methods, but the finishing becomes difficult, because surface E must be faced off when turning pilot A, and since interrupted turning operations should be avoided.

If a redesign to eliminate this condition will not adversely affect the end results, it should be considered.

and face surface E in one clamping. But, of course, in facing surface E there was a bad interrupted cut.

With a little thinking, you can change the design as shown in Figures 4.9-2 and 4.9-3. The part is still made of cast iron, and surface E can either be milled or ground and the hole bored. (Figure 4.9-2) The pilot can then be made from a cast iron drum and turned a little oversize, then cut to length and cut with a hacksaw as shown. (Figure 4.9-3) Cast iron pilot was chosen because it is considerably springy without heat treatment and if made oversize will fit snugly in the hole in the part and assume the roundness and accuracy of the hole.

You now have an economical design that can be produced more accurately. You have avoided the interrupted turning operation, and in an oddly shaped piece you can produce a hole more accurately than an external short pilot. With a separate pilot, the hole in the mating part can also have a very small chamfer.

Always make sure before you get too far along with your design

4.9 Machining Practices

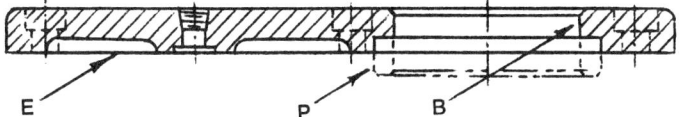

Figure 4.9-2. This shows a redesign of figure 4.9-1. Hole B can be bored and counter bored to receive pilot P. Surface E can be milled or ground.

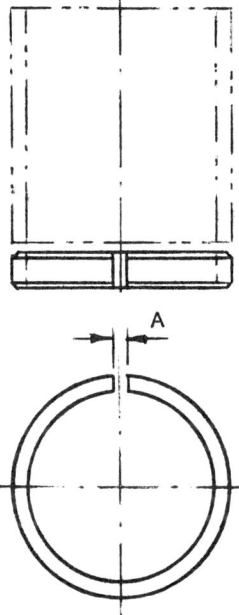

Figure 4.9-3. Cast iron pilot. Cut off from finished drum, saw cut and break corners before assemblying in counter bored hole in figure 4.9-2.

that your manufacturing machines can handle your design. You may have a ground threaded portion integral with a shaft, but the shaft may be too long for the grinder in your plant. It is best to find this out before the part is being detailed. Always consider stiffness of the part for manufacturing purposes also. It may be that your design will not make the part stiff enough for accurate manufacturing. In such case, provisions will have to be made by the process engineer for accurate manufacturing, which increases the cost. On the other hand, quite often the part can be stiffened, which will not, as a rule, impair the design in any way.

In fastening one part to another, especially two castings, it is well to provide for a mismatch, where the smaller part is about one sixteenth of an inch smaller all around than the larger part. This amount is, of course, varied in proportion to the size of the part and can be much less for two steel parts. This makes for a pleasing design and the parts do not have to be so accurately finished. (Figure 4.9–4)

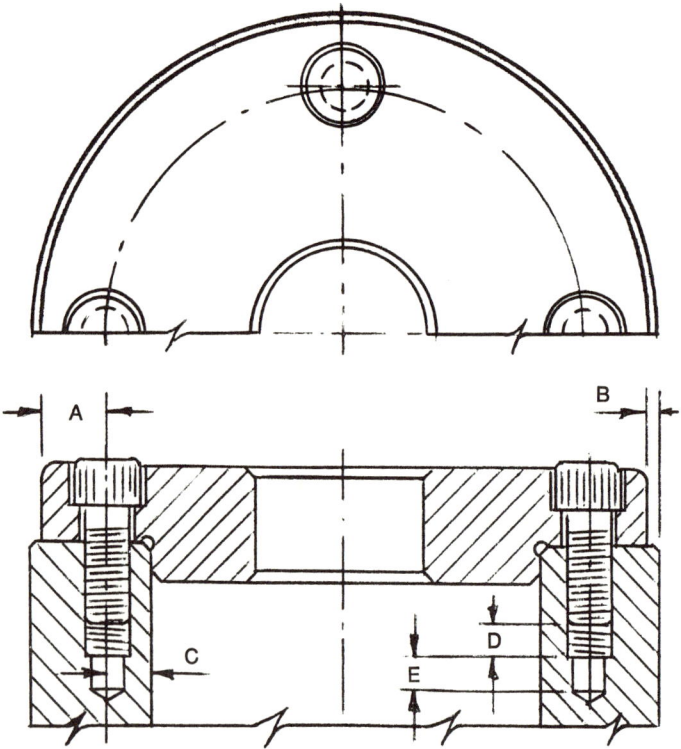

Figure 4.9–4. When there is a possibility that two surfaces joined together may not line up perfectly along the outside, it is better to provide for a small mismatch B. The inaccuracy then will not be as noticeable.

Follow consistent rules in placing screws from the edges. Practically speaking, a variation of a small amount would not make any difference as far as strength is concerned, but a pleasing design has exactly the same spacing throughout the machine within manufacturing limitations.

A good rule is to make distance A the same as the screw head diameter for rough castings, a little less for finished edges. Distance C should be the same as the screw diameter for general design, a little more for bearing housing bores and hydraulic purposes. When distance C is too small, the finished bore will swell up when tapping the screw hole. Dimension D should be of sufficient depth so that a plug tap can be used. Dimension E should be deep enough so that there is room for chips when tapping the screw hole.

4.9 Machining Practices

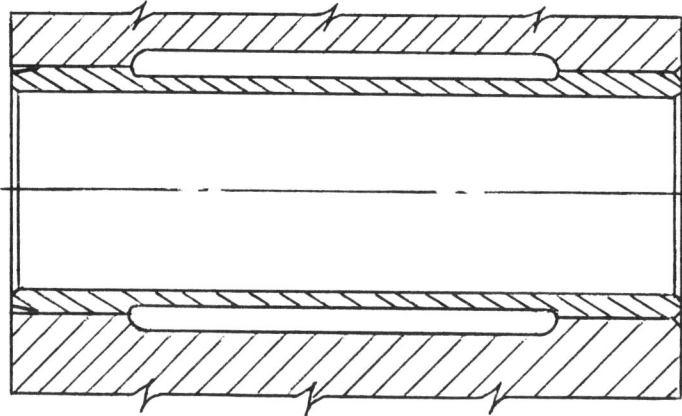

Figure 4.9–5. Layout showing economical design for a long hole receiving a sleeve, giving the highest degree of accuracy

When your design calls for a long sleeve in a hole, be sure to relieve at least one of the members and preferably both. (Figure 4.9–5) You then have the best control, for you are sure that you have a bearing at both ends. The parts are also much easier to finish.

The same holds true for two flat surfaces joined together. A narrow contact surface can be produced faster with better accuracy and gives the best control when the two parts are fastened together.

A layout should be accurately prepared to show the detailer how the parts should be finished. (Figure 4.9–6) This drawing shows the preparation of shaft, housing and sleeves for ball bearing mounting. All sharp corners should be broken, if possible. This may not be easy to show on the layout, but should be taken care of on the detail drawing. So that the seats may be finished accurately and square, all corners should be relieved, with proper radius in the relief.

Finishing Methods

Grinding is by far the most common method of finishing for good surface integrity. The machining method used to produce a surface and the parameters used in any machining method can have an important effect on the nature of the surface produced on a component. Since the surface condition often controls the basic properties required for successful application of parts in service, the surface must have an inherent quality, which has been referred to as surface integrity. All machining procedures tend to produce a surface layer somewhat dif-

Prototype Machine

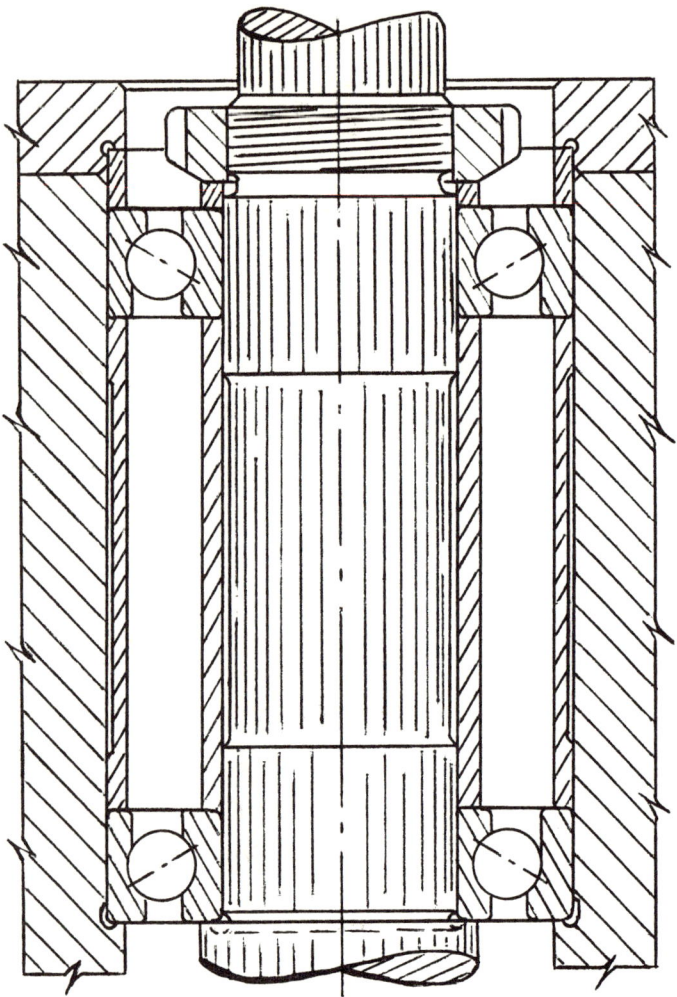

Figure 4.9-6. Accurately prepared layout ready for detailer

ferent from the matrix of the material. This surface layer may be shallow or deep, as a function of the severity of the machining operation. In general, the more abusive types of operations, that is, those which produce the greatest amount of plastic deformation or the highest temperatures, tend to produce the greatest change in the surface. Abusive machining and grinding tend to introduce high residual stresses, which may in turn result in large distortion of the workpiece.

High wheel speed and down feed in surface grinding tend to pro-

4.10 Fastening Practice and Proportioning

duce high residual stresses as well as distortion of a high magnitude, whereas gentle down feeds and low wheel speeds tend to produce low stresses and small distortions. In grinding, two additional parameters have significant effect on the distortion and residual stress. These are the type of grinding wheel and the grinding fluid. Soft wheels and chemically active cutting oils tend to minimize stresses and distortions.

Abusive machining tends to produce surface layers containing excessive plastic deformation and possibly metallurgical transformations. For example, on steels which are abusively ground or milled, a hard white layer of untempered martensite may be formed.

The surface changes produced in machining of components may subsequently affect the fatigue strength, stress corrosion, and other mechanical or physical properties of a material. The seriousness of this alteration is a function of the material being machined, the extent of the surface damage, and the environmental service and stress to which a component is subjected. It is wise to consider carefully the surface integrity of the part in planning a machining operation to assure that the surface produced is satisfactory for the material and its application.

4.10 Good Fastening Practice and Good Proportioning of Fastened Parts in Machine Tool Design

The most common means for fastening two machine parts together for machine tools are socket head machine screws. The machine designer's primary concerns are that the fastened parts are securely held together and that the fastening means do not destroy the appearance of the design.

Secure fastening is most important. There are many ways to make the fastening of parts secure. We shall consider a few of the basic reasons why parts become loose due to vibration. If the designer will keep them in mind as the design progresses, he can work out each case in a way most suitable to the design. Just to add a lock washer will not solve the problem, in fact it may, in some cases, make it worse and be a primary cause for failure.

Figure 4.10–1 shows two parts fastened together by means of a screw A and a lock washer B. Part C may be a cap or cover of minor importance. Stiffness has not been considered and the part may have warped slightly in the process of being manufactured. If made of cast iron, for instance, an unequal stress pattern is set up when the rough skin is removed on the one side of the casting and the part will warp.

Table 4.9–7.
Surface Roughness Produced by Common Production Methods *

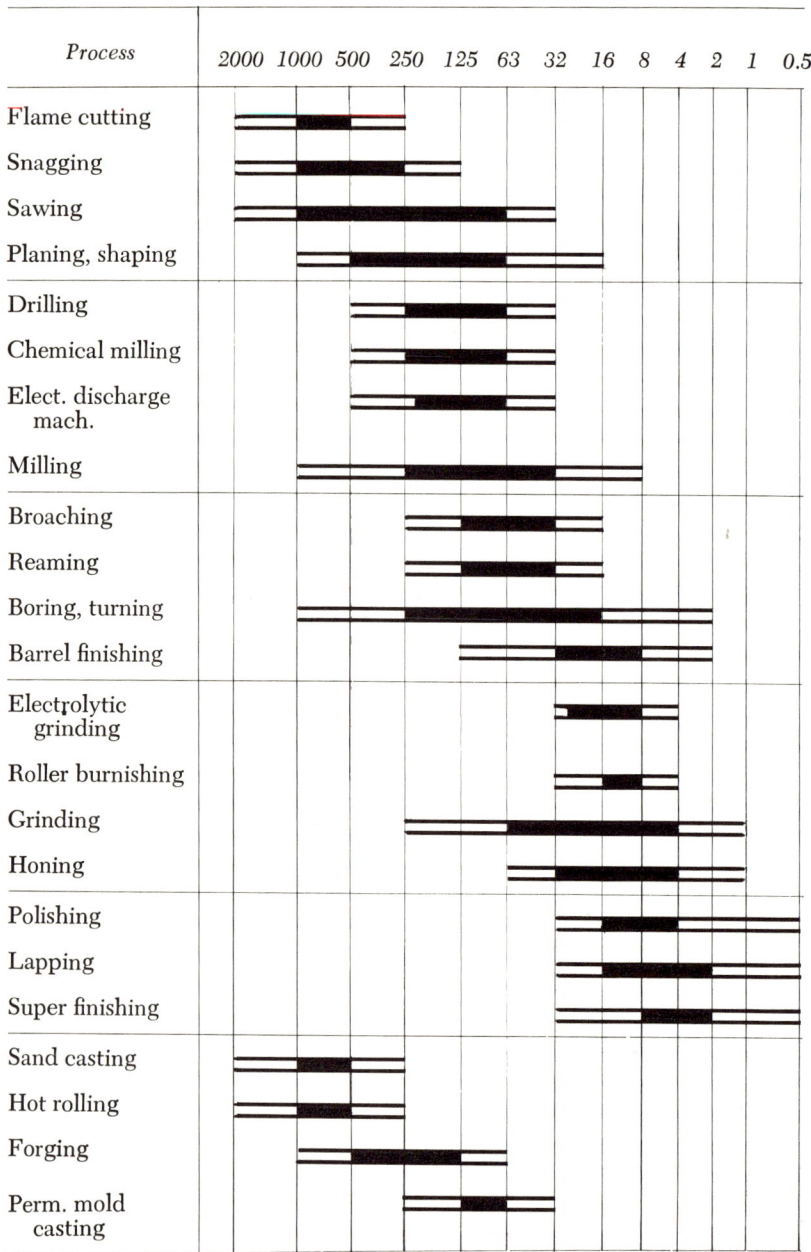

4.10 Fastening Practice and Proportioning

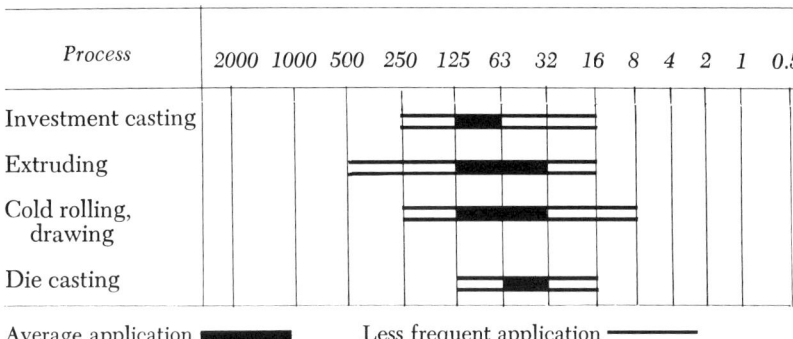

Process	2000	1000	500	250	125	63	32	16	8	4	2	1	0.5
Investment casting													
Extruding													
Cold rolling, drawing													
Die casting													

Average application ▬▬▬ Less frequent application ══

The ranges shown above are typical of processes listed. Higher or lower values may be obtained under special conditions.

* Extracted from USA Standard Surface Texture B46. 1–1962 with the permission of the publisher, The American Society of Mechanical Engineers, United Engineering Center, 345 East 47th Street, New York 10017.

Then, even if distance D is small when the screw A is tightened down, vibration originating in another part of the machine will cause the screw to come loose. If the gap between part C and E were closed completely, the two parts would vibrate as a unit with less likelihood of coming apart, but chances are that enough tightening force cannot be put on the screw A and you therefore have a gap D. You have added a spring lock washer B which only adds to your problem, because by tightening the screw, the energy stored up in the spring washer will be released when vibration in part C occurs. This will hasten the loosening of screw A. Therefore, for a permanent fastening, do not add a lock washer. Do make sure that your design will permit the surfaces of the two fastened parts to contact directly under the screw, so that if any vibrations take place the two parts will vibrate as a unit.

One way to accomplish this is to have a washer under each screw between the two surfaces A and B. These washers should be finished to exactly the same thickness. (Figure 4.10–2) This may be a necessary and only way of fastening two parts permanently and securely in some cases. The best practice in any case is to have as few parts as possible and the simplest design.

If one part can be relieved between the screws, this may be a good

Prototype Machine

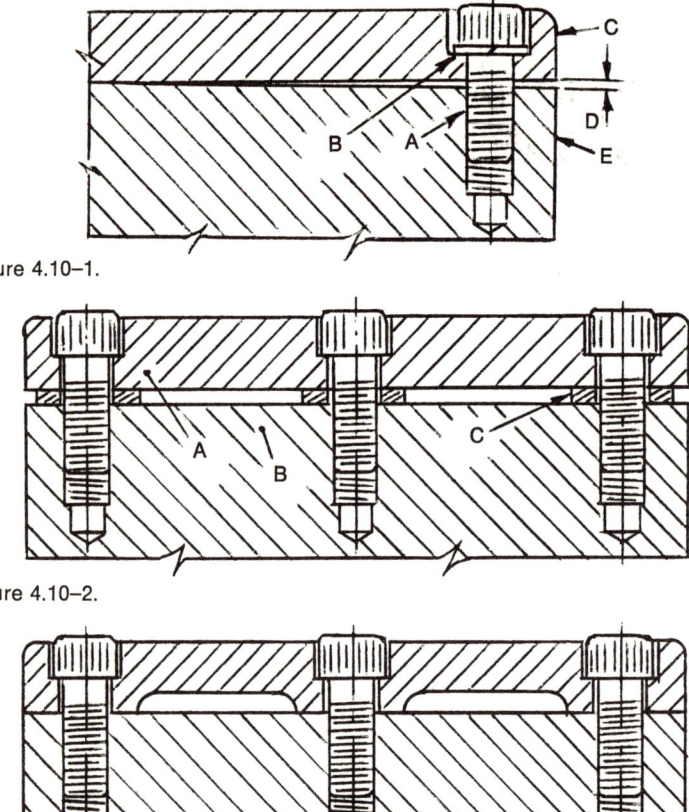

Figure 4.10–1.

Figure 4.10–2.

Figure 4.10–3.

solution. (Figure 4.10–3) In some cases, however, this may be undesirable and, therefore, another method is shown in Figure 4.10–4. Here you have ample contact surfaces around the screws and just a narrow surface between the screws to make a complete enclosure. Of course, each case has its own solution peculiar to the design under consideration. A steel part could not be economically relieved as shown in Figure 4.10–4, but for a casting, this would be an ideal arrangement because it would not cost more to produce the casting, and the machining would cost less. Warpage of the part from machining stresses would also be minimized.

4.10 Fastening Practice and Proportioning

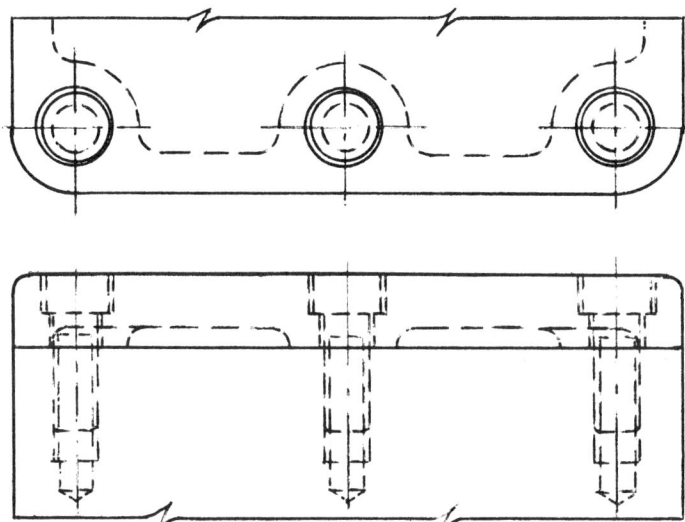

Figure 4.10-4.

The size of the screws should be chosen to meet the design requirements. If it is just a matter of secure fastening, a few screws of sufficient size for the load may be just right for the job. If other things must be considered, such as oil or air enclosures, several smaller screws, placed closer together, may be necessary. The final cost may be slightly more, but performance, quality and cost must be considered together.

Another thing to be considered for good fastening is the length of the fastening means, in the cases described, a screw. The fastening may take other shapes than a simple screw, but the same consideration is applicable. A screw can be considered as a spring, since there is no rigid material in engineering. Under a certain tightening pressure, the screw will elongate, the amount of elongation being a function of the length and diameter of the screw and the load and material used. A screw two inches long between the head of the screw and the contact surfaces, will stretch twice the amount of a screw of the same material one inch long between the same points and tightened to the same pressure. Therefore, the screws or other fastening means should be as short as possible, considering the strength of the fastened part. Figure 4.10–5 shows an undesirable condition, the screw being too long. A much better design is shown in Figure 4.10–6, using a short screw. The counterbored hole can then be plugged with a simple welch plug, or,

Prototype Machine

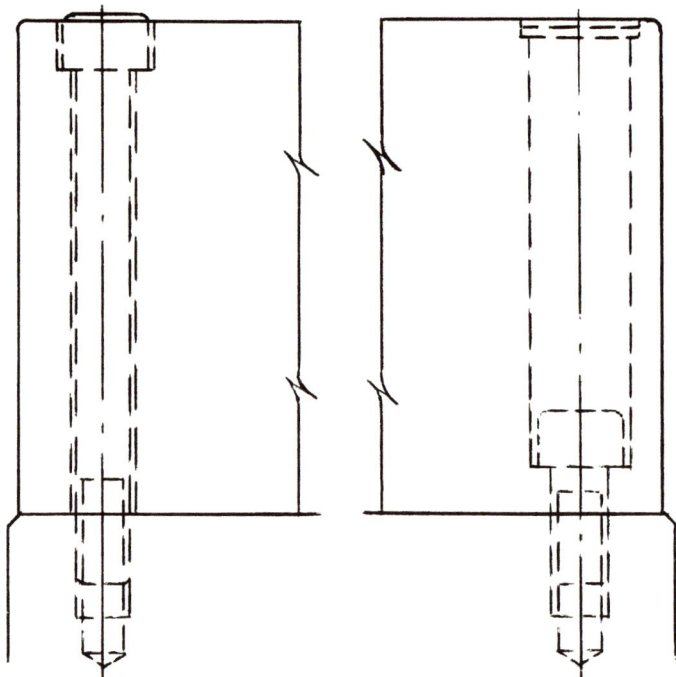

Figure 4.10-5 (left) Figure 4.10-6 (right)

if you have a finished surface, with a cast iron plug. A plastic plug may also be considered.

If you have chosen screw sizes to meet strength and other important considerations, the material fastened and the material fastened to should also be designed to meet these requirements, and to have a pleasing relationship to the screws. Standard proportions can then be worked out which make the best arrangement for appearance. There may be many load conditions that determine thickness and ribbing. A good designer always works for a pleasing appearance, and as a rule, this is also the best arrangement. The diameter of the screw in a particular machine component usually works out to be that based on the best proportion for distance from center of screw to edge or wall. (Figure 4.9-4) Whatever proportion is adopted should be followed throughout the design.

If the fastened part has a certain close relationship to the part it is fastened to, this relationship must be maintained by the use of a key, pilot or dowels or a combination of these. If, however, no relationship

4.10 Fastening Practice and Proportioning

is required, the screws alone will do a satisfactory job.

There are cases where a flush job is required in fastening sheet metal parts or relatively thin finished parts. This can be done with countersunk screws. (Figure 4.10–7) Make sure that this arrangement is absolutely necessary. In many cases a low head cap screw will do the job. The countersunk holes in the fastened part can be manufactured to comparatively close tolerances as far as spacing is concerned. The part fastened to, however, has threaded holes for the screws, and it is impractical to thread several holes to close spacing accuracy. In manufacturing the screws, the conical surface of the head cannot be

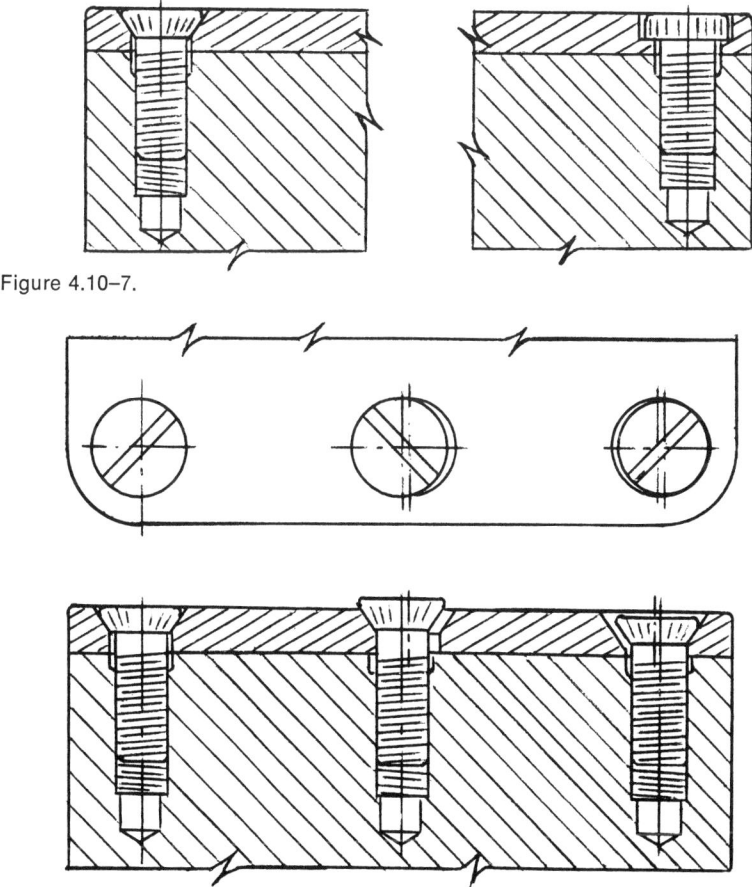

Figure 4.10–7.

Figure 4.10–8.

Prototype Machine

held to a very close manufacturing tolerance in relation to the threaded portion of the screws. When the parts are assembled, you have a very unsightly condition as shown in Figure 4.10–8. This is also not so good from the standpoint of security in fastening, for you have just a line contact between the screw head and the fastened part.

A bearing cap usually has a short pilot to center it to the bearing housing, but quite often, it also has a key or a dowel to locate it angularly, if angular location must be accurate. To locate an oil hole or groove in the bearing cap to an oil hole in the bearing housing, displacing one of the screw holes 5 or 10 degrees angularly is close

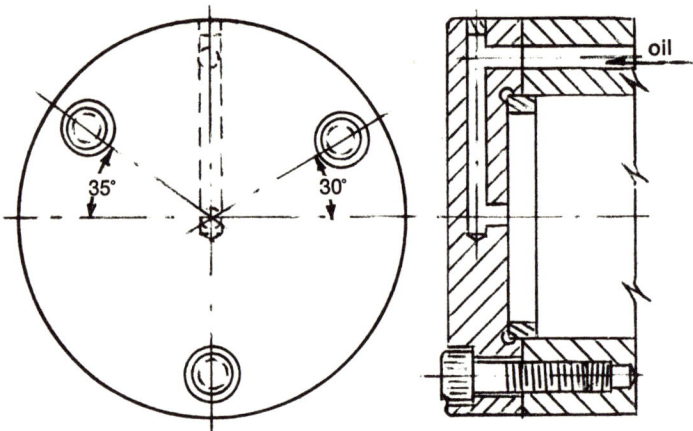

Figure 4.10–9.

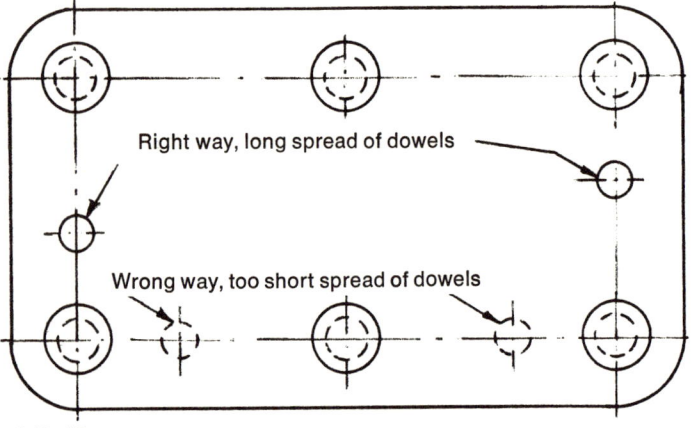

Figure 4.10–10.

4.10 Fastening Practice and Proportioning

enough. This will assure you that the bearing cap can only be assembled in one angular position, and you have not added anything to the cost of the design for this feature, as shown in Figure 4.10-9.

In some cases, you may want to locate an oddly shaped machine part by dowels as shown in Figure 4.10-10. The dowels should then be located as far apart as possible to give the best control. A couple of soft steel dowels may be used if this is a permanent fastening. If, however, this should be a very accurate fastening, and the part has to be removed frequently, a couple of hardened dowels should be used, preferably pull dowels, e.g. dowels having a threaded hole in one end

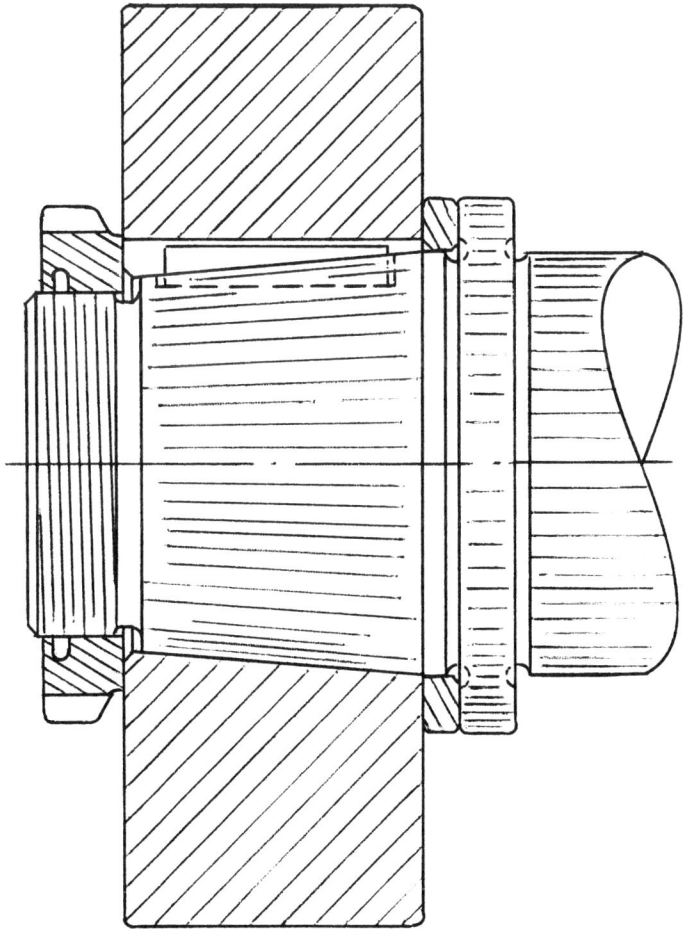

Figure 4.10-11.

Prototype Machine

so that a threaded rod can be inserted to pull the dowel out.

Cams, cutters and other important parts that must be fastened accurately are usually centered with a tapered bore. Standard bore sizes should be adhered to for interchangeability. The part can be located angularly with a key. If the part is fastened only at the center, care must be taken to draw the part down with the same pressure each time, so that the bore is not unequally expanded, which would cause a variable distortion. A simple way to do this is to provide the spindle with a small flange a short distance back of the part. The part can be fastened the first time with a torque wrench. When using a torque wrench, the surfaces should be dry because oil or grease will give false readings. The distance between the part and the flange can then be measured accurately and a washer can be ground exactly to the measured dimension. With the washer between the part and the flange, any subsequent fastening of the part will be accurately repeated with no variable distortion of the part, as shown in Figure 4.10–11, and dry or oily surfaces will not have to be considered.

If the part is a large diameter cutter, the tapered hole is more likely to be used just for centering and the fastening screws are probably on a large diameter. The best arrangement then is to use a torque wrench for tightening down the screws. There is a great difference in force applied when you have a dry or a lubricated screw, so this must be considered.

Spur gears, worm gears and small bevel gears are usually fastened with a central nut. For accurate results, the screw threads are then ground. Permanent relationship to the shaft is usually maintained with a key.

Large bevel gears and hypoid gears are commonly manufactured as ring gears and fastened to a hub which then is centered on and fastened to the shaft or spindle with a central nut. This simplifies the entire method of manufacturing, because the gear can be made of a uniform section that is much easier to mount for manufacturing and much easier to handle for heat treatment. The gear would be hardened and ground steel and the hub could be fabricated steel or cast iron. The gear is centered to a short pilot on the hub and fastened with a number of uniformly spaced screws. The screws should have fine threads, and no washers are required under the screws. No dowels are used in high precision gears for this would mean a non-uniform section and thus be cause for distortion of the teeth. The friction between the

4.10 Fastening Practice and Proportioning

gear and the hub is sufficient to maintain the relationship for the maximum hp the gear is capable of transmitting. The screws should be locked, especially where there is a possibility of reversing, shock or

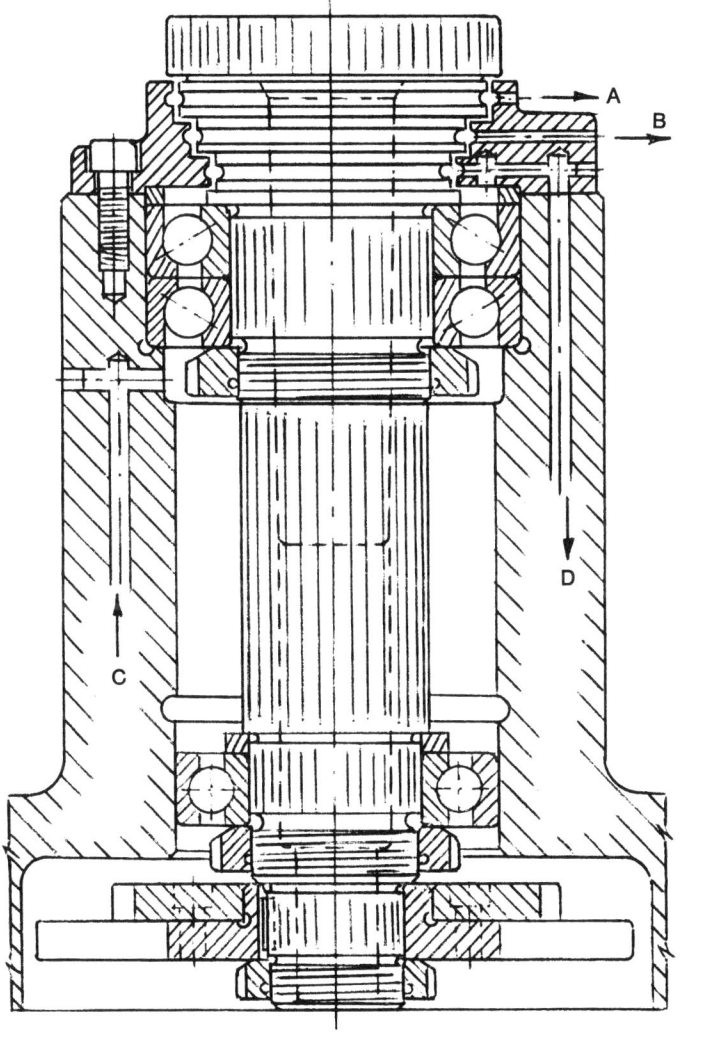

Figure 4.11–1. A work piece spindle similar to figure 3.5–2. A pair of preloaded ball bearings at the front of the spindle take the radial and thrust loads originating at the work piece. The single radial bearing at the rear of the spindle takes the radial load from the drive components.
   Coolant is drained off at A and B, so it may not enter the bearing.
   Lubricant enters at C and is drained back to the sump at D.

intermittent heavy loads. A small dowel through the head of the screw swedged in place is usually good.

Use washers under screws and nuts only when tightening against

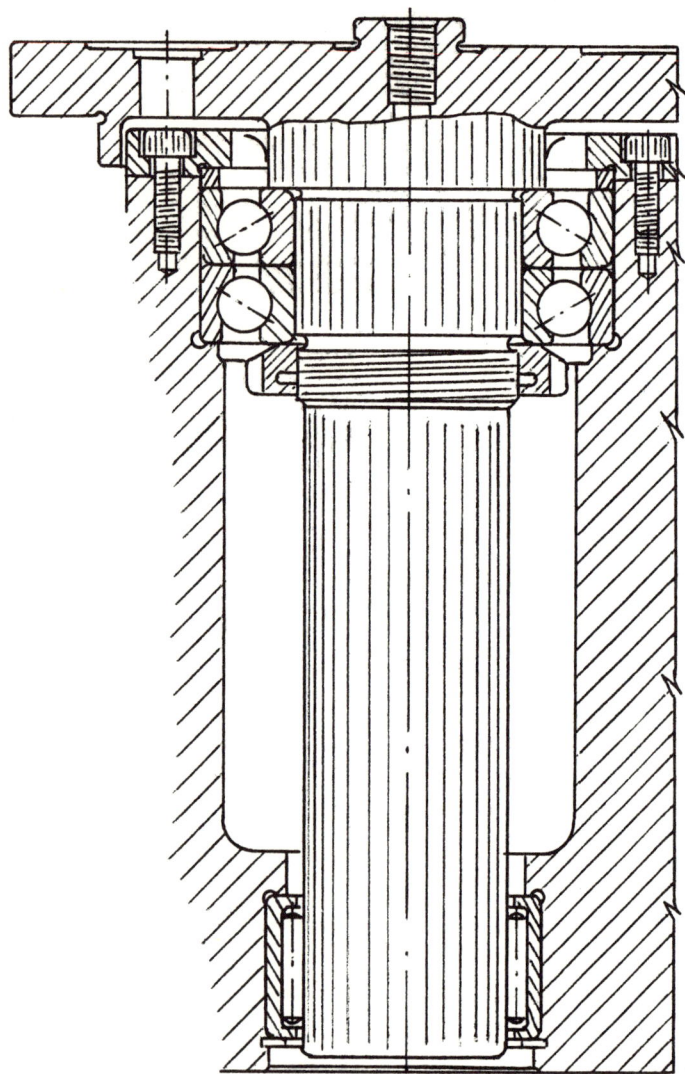

Figure 4.11–2. A cutter spindle similar to figure 3.5–1. A pair of preloaded ball bearings take the radial and thrust loads from the cutter and the drive gear. See figure 3.5–1. The roller bushing at the rear of the spindle stabilizes the spindle and takes a small part of the radial load imposed on the front bearings.

4.11 Main Spindle Layouts

extremely soft materials, when fastening parts which must be removed frequently or if a sleeve dowel is used with the screw for definite location. Use lock washers only when no other methods are possible.

Always remember that when two or more parts are fastened together and one or more of the major parts are subjected to a bending moment in the process of being fastened, something is lacking in the design, and this can not be improved by the addition of a lock washer.

4.11 Main Spindle Layouts for Machine Tool Design

We shall mention a few examples of kinds of spindles from which you can draw a conclusion for the best approach to a successful solution to your design problem.

For the ordinary spindles where speeds are low, (Figure 3.5–1) or for intermittent motion with high speed partial rotation of very short duration and comparatively long periods between rotations, (Figure 3.5–2) a straight forward solution is possible with simple bearing mountings.

A simple solution for ball bearing mountings is shown in Figures 4.11–1 and 2. These mountings will also take care of heat expansion for higher speeds, because the rear (floating) bearing can move axially in the housing, or the (floating) spindle can move in relation to the bearing.

There are cases where the solution is not so simple; for instance, with very high speed spindles. When a spindle is started up to attain a high speed, heat is generated rapidly in the revolving parts, even with anti-friction bearings. It would seem that a sufficient supply of lubricant would remedy this situation. This is not so, because the high speed churning of the lubricant only adds to heat generation. The inner bearing race and the rolling components heat up first and a rapid radial expansion of those parts is the result. Heat transfer is not instantaneous and since time is required for heat to be transferred to the outer parts and released to the surroundings, the preload of the bearings is rapidly increased. Before the outer parts have had time to expand sufficiently, the bearing may have suffered irreparably in spite of good lubrication.

One simple solution to this problem is the use of tapered roller bearings. Some examples are shown in Figures 4.11–3 to 4.11–6.

In high speed spindles, generated heat due to high speed is not the only difficulty. The system must also be designed to provide a good

Prototype Machine

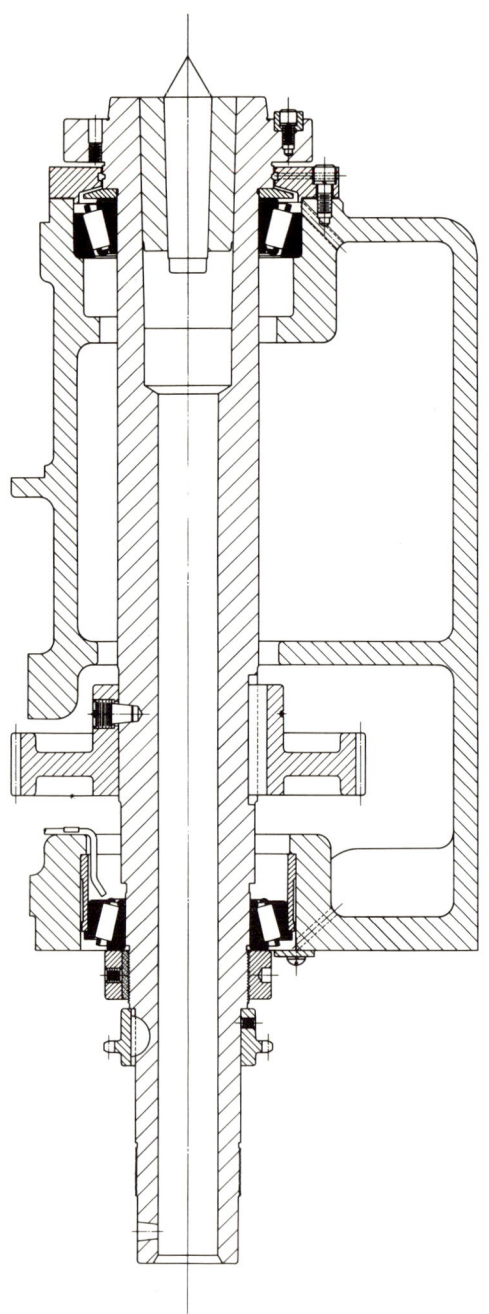

230

4.11 Main Spindle Layouts

safety margin between the critical speed of the spindle and the actual operating speed. Where space is available, this may be comparatively easy to work out. Forces transmitted from the driving components may sometimes also be troublesome. In Figure 4.11–7 and 4.11–8 some solutions are worked out. Here the spindles are not solidly connected to the driving source. Therefore, only torsional deflection is induced on the spindle from the driving source.

Springs are sometimes used for preloading the bearings on high speed spindles. Figure 4.11–7 shows a unique way of preloading the bearing, using air pressure. Air is available in most plants and is an economical method since no air is lost in the process. If different parts of the machine cycle call for variable preload, this can be easily taken care of with air.

In planning the design of a spindle, it should be remembered that outside sources also tend to build up heat in a bearing. The spindle has been made heavy enough for the forces involved and long enough to provide for stability. These conditions will help to absorb the heat, but also remember that a large, heavy flange at the front of the spindle will help to dissipate the heat fast to the surroundings.

Always keep in mind that loss of preload in the bearings due to heat expansion is one of the main causes for chatter marks in the part being processed, whether this loss of preload occurs in the work piece spindle or the grinding wheel or tool spindle.

For some reason it may be necessary or desirable to have a drive

Figure 4.11–3. Preloaded high speed spindle for tapered roller bearings. Because the bearings are mounted far apart on a long spindle, and heat transfer requires time, the bearing cones and rollers will expand radially long before any axial expansion has taken place in the spindle, which would result in excessive tightness of the bearings soon after they have attained high speed. Since the heat is slowly transferred to the spindle, this axial expansion would eventually cause looseness in the bearing mounting. Therefore, the bearings must be preloaded sufficiently to eliminate looseness for high speed running conditions.

To remedy excessive tightness, the rear bearing cup is mounted in a sleeve adapter pressed in the housing. As shown, a short land on the sleeve in front of the bearing cup is made a light press fit. The diameter of the short land on the sleeve directly outside the cup is made sufficiently smaller in diameter than the housing bore to allow for radial expansion when the bearing heats up. The sleeve diameter between the two lands should be relieved and have generous radius at each land. A flanged cup is used for the rear bearing to make the manufacturing of the sleeve simple. (Courtesy of The Timken Roller Bearing Company)

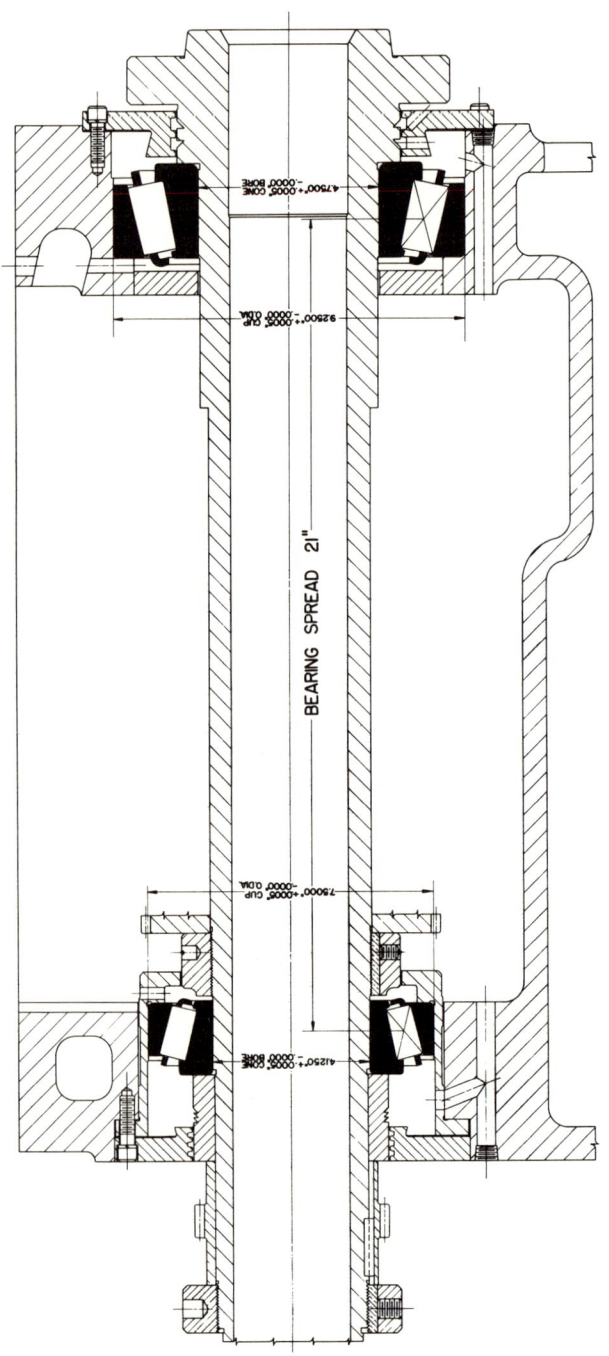

Figure 4.11-4. This shows a high speed spindle mounting for tapered roller bearings using a standard cup. The flange of the sleeve may be just enough to back up the cup, thus eliminating undesirable stiffness. (Courtesy of The Timken Roller Bearing Company)

4.11 Main Spindle Layouts

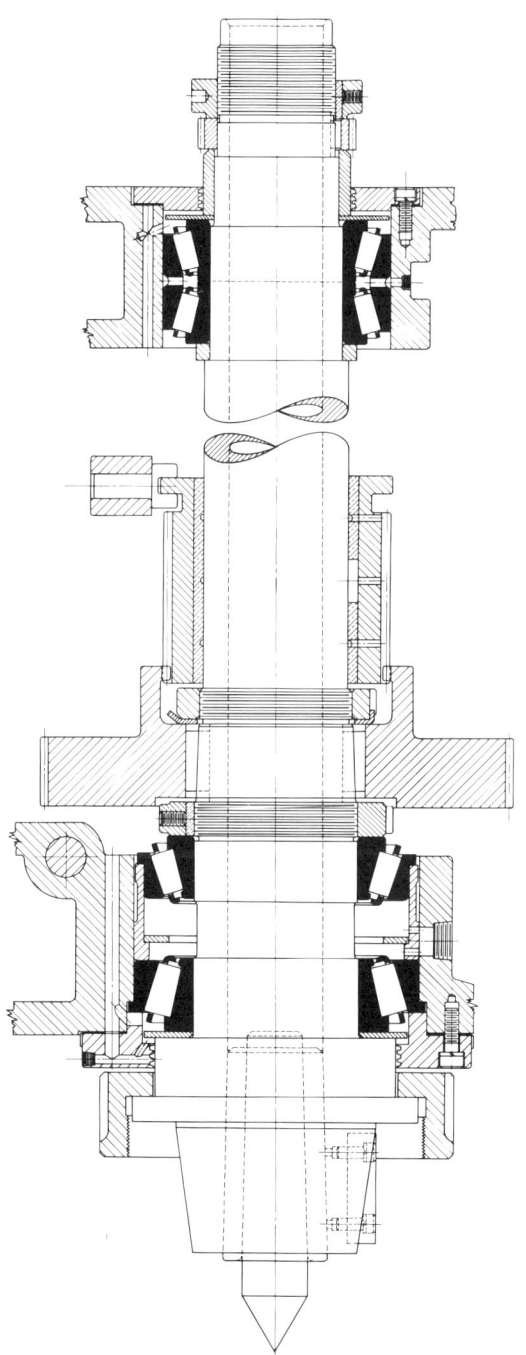

Figure 4.11-5. High speed, heavy duty production lathe spindle for tapered roller bearings. The expansion sleeve is here used at the rear of the two front bearings. (Courtesy of The Timken Roller Bearing Company)

Prototype Machine

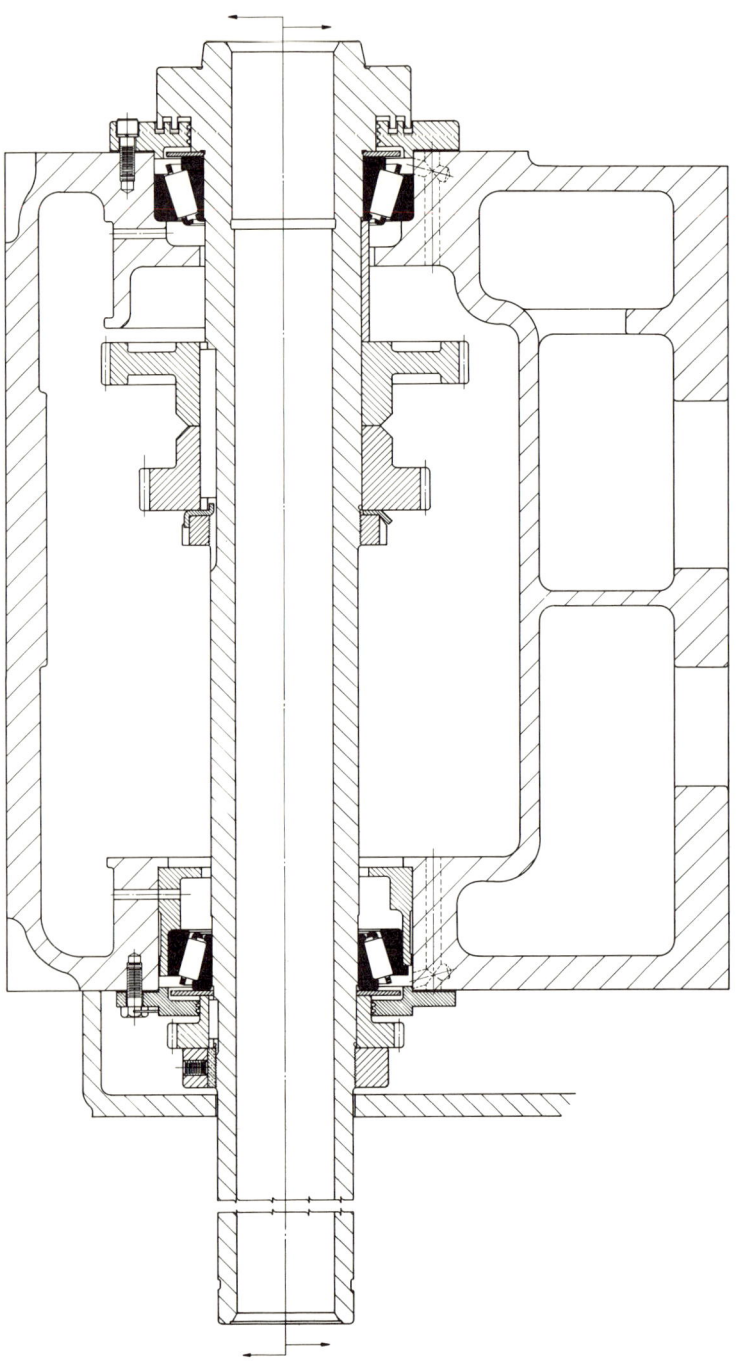

4.12 Main Drive Assemblies

gear or pulley, or a combination of these components on the main spindle. The tooth separating forces or radial pulley load would then put additional radial and thrust loads on the spindle. A simple way to eliminate these additional loads from the spindle is shown in Figure 4.11–8. Many variations of this design may be worked out to satisfy a particular design.

4.12 Main Drive Assemblies for Machine Tool Design

The main part of the drive is usually arranged in a compact housing well ribbed to support the shafts carrying the mechanism. Thought should be given to the location of this housing, so that all replaceable parts such as cams and change gears are easily accessible. It should be arranged so that it is a self-contained unit that can be assembled and tried before being joined to the rest of the machine. This main drive assembly housing should also be arranged so that it is properly fastened to the main frame of the machine. If the housing must be lined up with another main part of the machine, this can be done by having a shim between the housing and the frame. This shim could be made of cast iron and should be at least one quarter inch thick when finished. Thinner material is difficult to keep flat and parallel when it is being finished. The shim should be stocked a little oversize, say five sixteenth inch thick.

Using shims is the best method when it comes to assembling large assemblies to be joined to a high degree of accuracy. Perhaps other minor assemblies also must be fastened precisely to the main drive assembly. If there is only one center to be lined up to, this can usually be taken care of with a short pilot on the auxiliary assembly. If there are several centers to be lined up to, and you cannot fasten the auxiliary assembly to the frame of the machine, this can usually be assembled to the main drive assembly by providing a finished ledge on the main drive bracket a little below the auxiliary bracket. You can then have a shim on this ledge with the auxiliary assembly resting on the shim. This takes care of the vertical alignment. The horizontal alignment is usually taken care of with a dowel. (Figure 4.12–1)

Figure 4.11–6. Turret lathe high speed spindle. A standard cup is shown above the center line and a flanged cup below the center line for a variation of mounting for the rear bearing. (Courtesy of The Timken Roller Bearing Company)

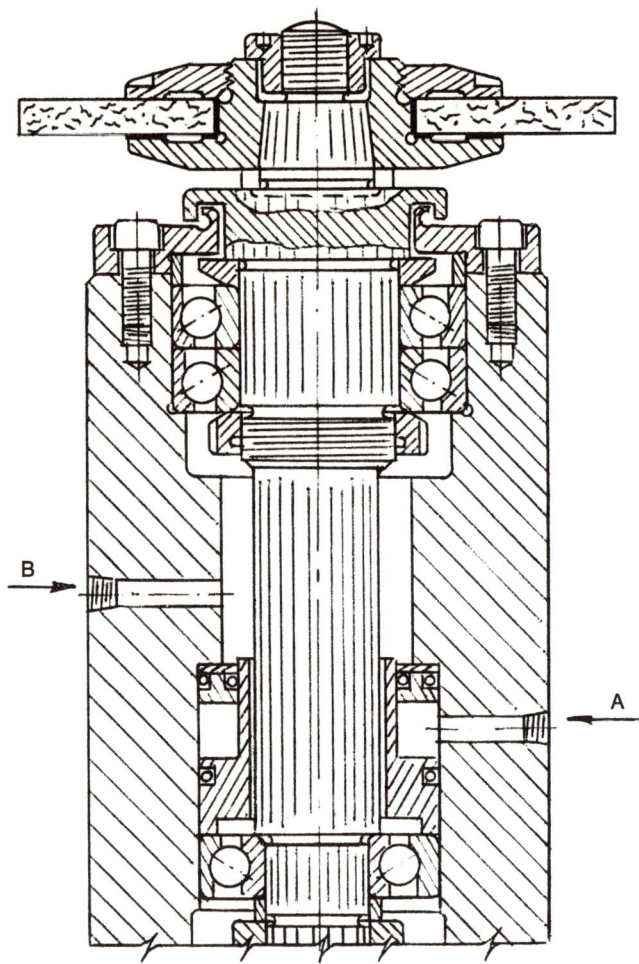

Figure 4.11–7a and 4.11–7b. High speed spindle mounting. This is shown for a grinding wheel, but may also be used for a high speed mill spindle using tungsten carbide blades or a similar arrangement.

Two high angle ball bearings are used in front of the spindle. These bearings, which carry all the thrust load and most of the radial load, are preloaded with a single bearing at the rear of the spindle. The preloading medium, entering at A may be air or oil, whichever is available.

The drive spindle is not solidly connected to the main spindle. At C is shown a Gleason CURVIC* coupling with zero pressure angle. This coupling is manufactured with spiral bevel gear machines to a high degree of accuracy. A small axial motion, due to heat expansion, is accordingly provided for and only torsional deflection is imposed on the main spindle from the driving source.

This design may be helpful considering the critical speed of the spindle, since this is a function of the bending deflection at the center of the spindle.

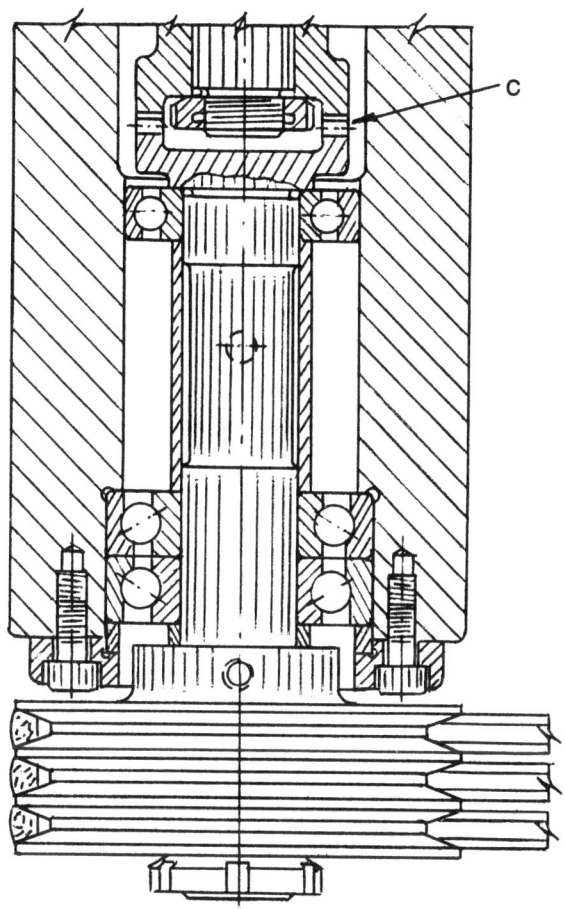

$$y_o = \frac{W \times 1^3}{48 \times E \times I} = \text{Bending deflection at center of freely supported spindle}$$

$I = .049 \times D^4$ for solid shaft
$I = .049 \times (D^4 - d^4)$ for hollow shaft
W = Load in pounds
1 = Distance between bearings in inches
$E = 29,000,000$ for steel
$E = 10,000,000$ for Duraluminum
$E = 6,500,000$ for magnesium

*CURVIC is a registered trademark of The Gleason Works

$$N_c = 187.7 \sqrt{\frac{1}{y_o}} \quad \text{Critical speed in rpm}$$

Disturbance originating in the drive components are not transmitted to the main spindle, as when the drive is solidly connected to the spindle.

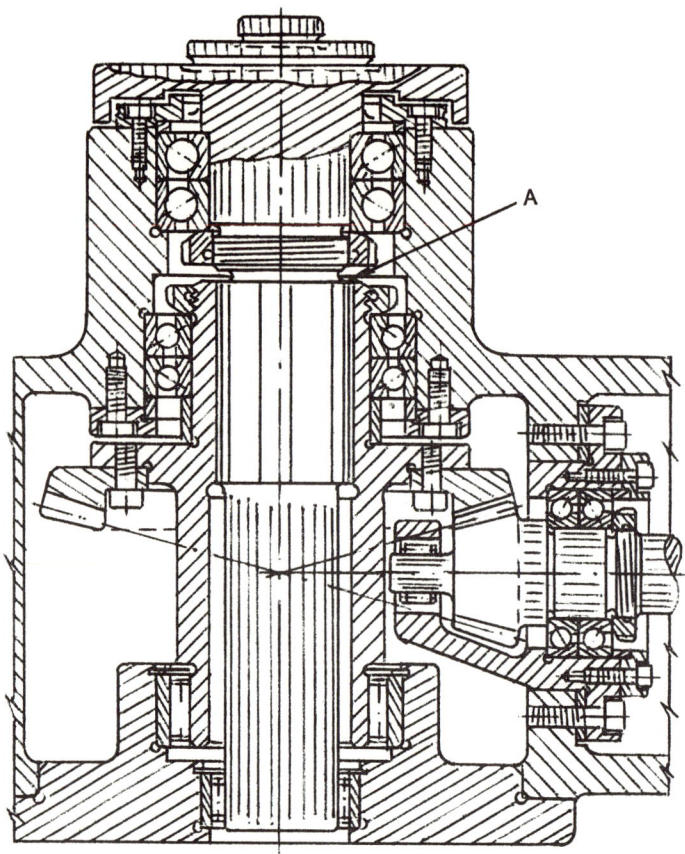

Figure 4.11-8. This is a medium speed milling spindle. Here there is no rigid connection between the main spindle and the drive components. Therefore, the radial and thrust loads resulting from the tooth separating forces are not transmitted to the main spindle. The drive gear is mounted on a sleeve surrounding the spindle, and is supported by two preloaded ball bearings at one end near the gear, and a roller bearing at the other end. There is radial clearance between the main spindle and the inside of the sleeve. The main spindle is mounted on two preloaded ball bearings at front and a roller bearing at the rear. The load from the mill is close to the front bearing so only a small portion of the radial load falls on the roller bearing. The spindle is connected to the sleeve by three splines A and can, therefore, be dynamically balanced.

In the process of manufacturing, it is much easier to maintain a close relationship among several holes than to maintain a close relationship from one or two surfaces to several holes. Make a close study of assembly procedures as you work on your design so there is no alignment problem in any direction.

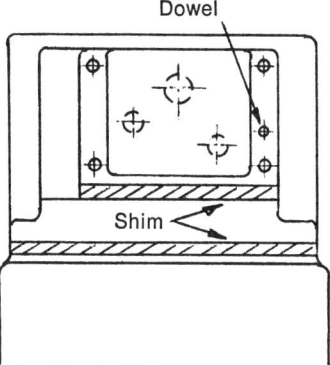

Figure 4.12–1a.

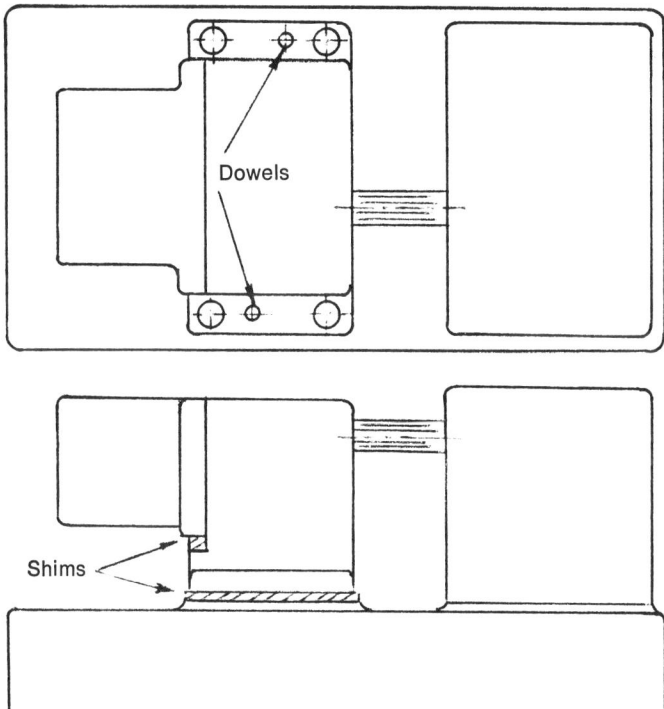

Figure 4.12–1b.

When assembling a part for accurate alignment to another part, the control for squareness should be closely considered. For a complicated design where so many other necessary requirements have been taken care of, such as stiffness to minimize deflection, control for squareness may not be as evident.

Prototype Machine

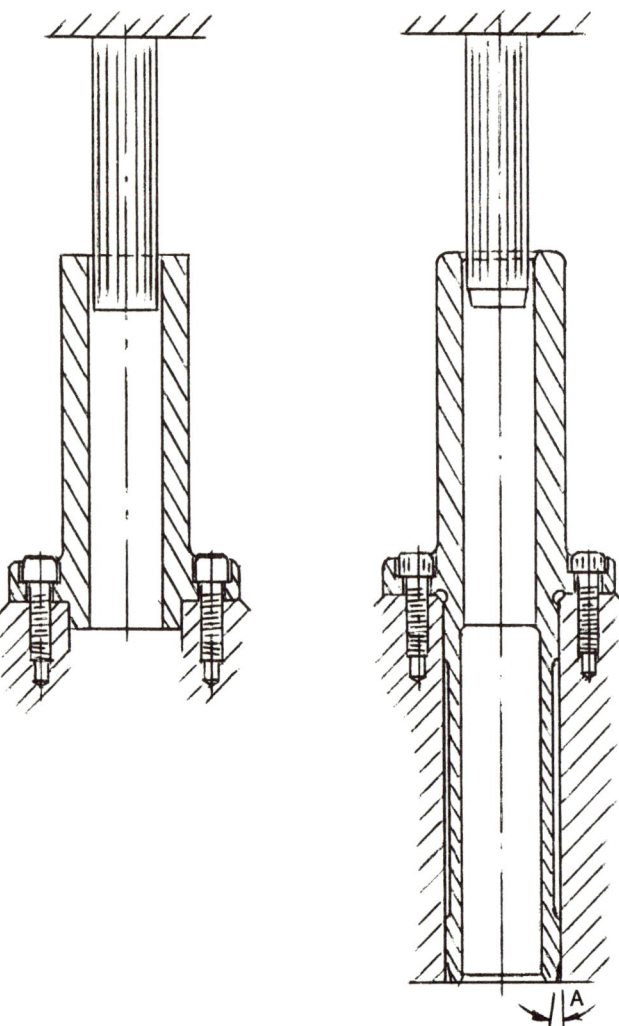

A circular part has been chosen to simplify the illustration, but it may also be applied to other shapes. Just to provide for positive location, as shown in Figure 4.12–2, is not enough. Control for squareness can either be provided for in the plane of location, as shown in Figure 4.12–3, or 90° to the plane of location, as shown in Figure 4.12–4. Never provide for squareness control in two planes, because in some cases, one will fight the other.

4.12 Main Drive Assemblies

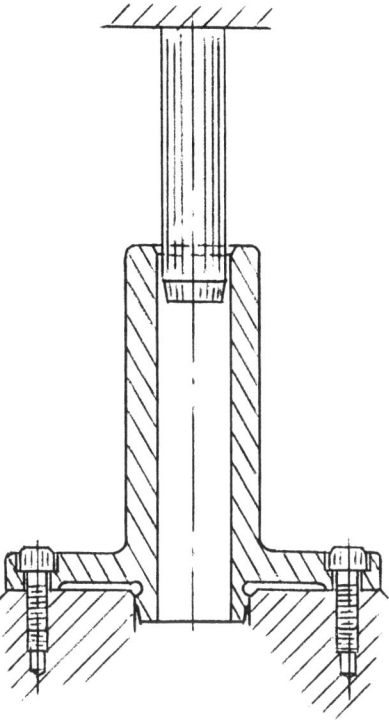

Figures 4.12-2, 4.12-3 and 4.12-4.

These drawings have been made in the simplest form to illustrate the necessity of good design for proper alignment. In figure 4.12-2 the short pilot would center the part. There are, however, no broken corners, so it would be difficult to assemble the part, and with the small flange the sharp corners would also make the alignment questionable.

In figure 4.12-3 the part is well centered and aligned with the long pilot which has a generous angular lead A. The pilot also has a short land at each end, all contributing to easy assembly. Furthermore, the hole is chamfered and the corner of the pilot is relieved at the flange.

In figure 4.12-4 the part is centered with the short pilot, which has an angular lead, and is relieved at the corner of the flange. The part is squared up with a flange at least as large as the part is long. Good alignment is assured by the flange being relieved almost out to the screws. If the flange has been faced off in the same setup as for turning the pilot and boring the hole, any little inaccuracy of the squareness of the flange will not affect the accuracy of the alignment.

These drawings have been shown only to stimulate thinking with respect to proper alignment of more complicated parts, when you also must consider calculations of deflection under load.

Change gears for varying the speed of a certain section of the drive are usually centered and squared on the bore. An angular lead should always be provided for on the male member for easy assembly. If you have a reversing drive, where you depend on great accuracy in repetition, do not tighten down the gear, but rather tighten against a shoulder on the shaft, leaving a few thousandths of an inch clearance. This then leaves the gear free to bear against the key when reversing the drive direction, assuring accurate repetition. If the gear were tightened down, it would perhaps bear against one side of the key just before reversing. Then when the reversing starts, and if the loads are light, the gear would still remain in this position, but if the loads should increase, the gear may suddenly slip over to the other side of the key because it is tightened down close to the center where you have a great mechanical disadvantage to the torque applied on a much larger radius.

4.13 Main Layouts in Machine Tool Design

An adequate number of layouts must be made to reveal all details to all interested parties. If there are any defects in the design, and the layouts have been made clear and revealing, these faults will show up before detailing starts. Complicated layouts should be sectioned with simple single section lines to show strength, support, assembly methods and operational details. There is no time lost in this extra work because the designer or layout man needs time to study the layout thoroughly, and since simple sectioning does not require mental effort, it can be done as the layouts are scrutinized for defects.

The machine was divided into several separate assemblies when the preliminary layouts were made, and you have a list of these assemblies. This list may be valuable in planning for production and assembly procedures. If the assembly is complicated, it probably requires several sectional views. It may also be of value to show part of the adjoining assemblies to establish relationship. Layouts of the whole machine may also be desirable. Show dimensions where accuracy of location is important. In the progress of the layout, you should also thoroughly study assembly procedures to make sure that no impossibilities exist. In complicated cases, especially where a certain sequence in assembly methods is required, write down your conclusions and print your suggestions on the layout. It should be noted that these are only suggestions, for in the process of assembly better methods may be

4.14 Lubrication

discovered. If definite instructions are necessary for good reasons, it should be clearly stated.

A typical assembly is shown in Figure 4.13–1.

4.14 Lubrication of Machine Tools

The human element still has to be considered in the upkeep and care of machine tools, even if automatic lubrication is provided for, so every precaution should be taken to guard against failure due to lack of lubrication. Where rapid failure would occur due to lack of lubrication, install automatic warning signals, or better still, pressure switches, which would stop the machine before irreparable damage could happen.

There are many lubricating components and systems that can be purchased, and effectively applied. All along the process of designing your machine you must think of lubrication and how to provide for it most effectively. Whenever dependence on the human element can be eliminated, it should be done. There are many hidden places in a machine tool that cannot be effectively checked. Accordingly, safety devices must be provided to keep the main oil supply in operation, and the individual revolving and sliding parts should be so designed that they receive ample lubrication automatically. Take advantage of the motion already in the machine and a simple, effective arrangement can always be worked out. Avoid ways that add complicated parts to the design, which would contribute to noise and trouble. In some cases, the configuration of the parts to be lubricated can be taken advantage of. A tapered roller bearing, for instance, will draw oil through the bearing by centrifugal force when oil is applied at the small end. There are many cases such as this and you should figure out the best way to suit each case. In many instances, a sliding part will provide a good pumping action, just by changing the part and surroundings slightly, or by adding a couple of ball check valves. These valves can be purchased and can also be very easily made at your own plant. (Figure 4.14–1)

Rod C is a necessary part of the machine designed for reciprocating motion. By making a small addition to your design, you can take advantage of this motion, making a simple pump for supplying splash or cascade lubrication to other parts in the machine. When rod C moves in direction B, oil is drawn up through check valve D to fill

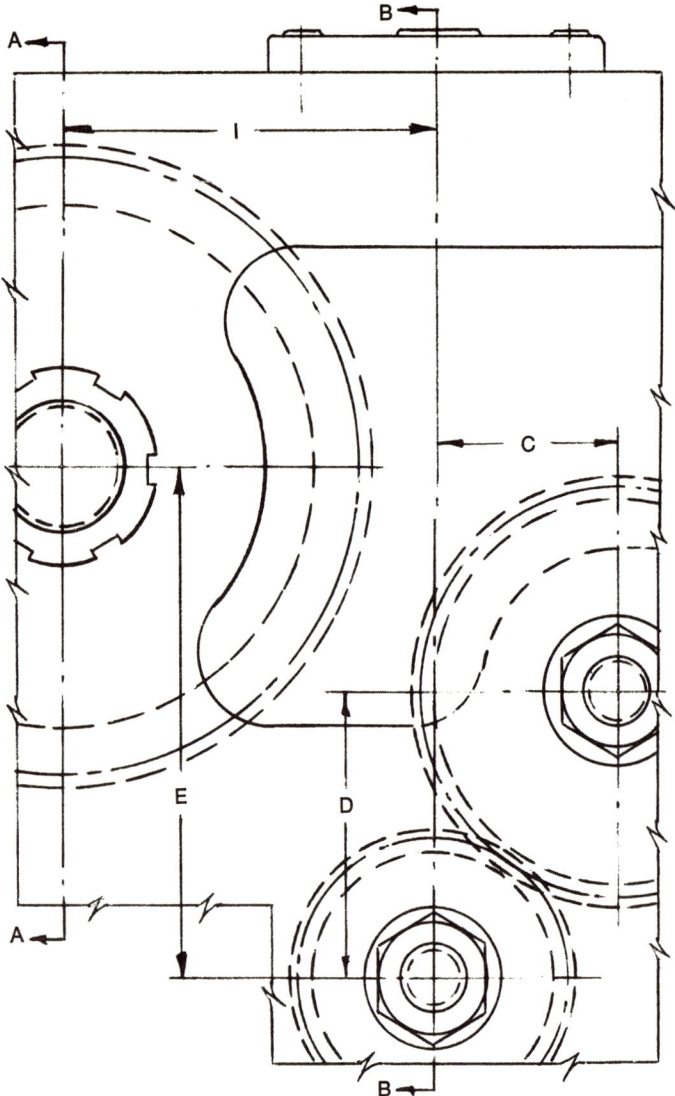

Figures 4.13–1a and 4.13–1b show a typical assembly layout. Important dimensions C, D, E and I are shown on figure 4.13–1a.

One bearing at each end of the shafts would not be good design, since there is no uniform section of support. See figure 4.13–1b. If one bearing were used, a cramping condition would exist when tightening the bearings for preload, due to the nonuniform sections, especially for the worm shaft for which the support section cannot continue around the entire circumference of the shaft on account of the worm gear.

4.14 Lubrication

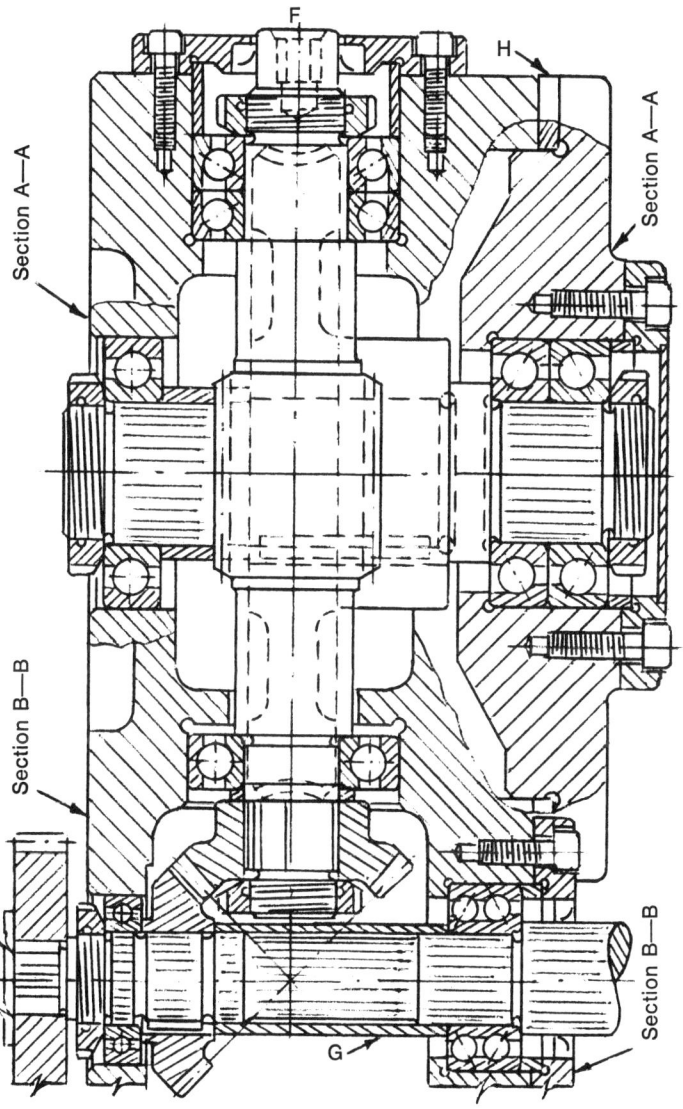

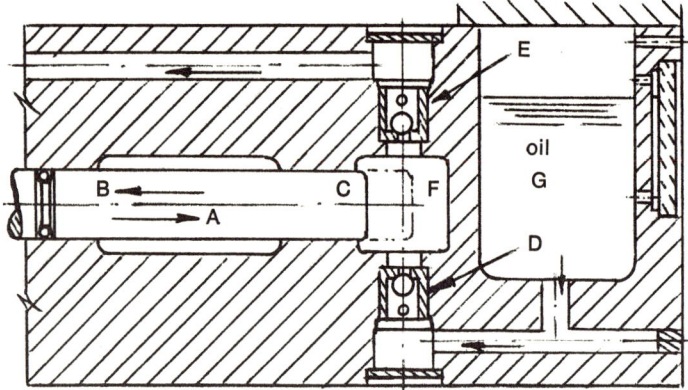

Figure 4.14–1.

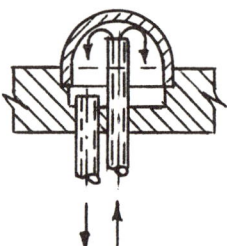

Figure 4.14–2.

chamber F. Check valve E prevents oil from leaking back from above. On the return stroke in direction A, check valve D prevents the oil from returning to the oil supply sump G. The oil trapped at F will, therefore, be pushed up through check valve E. By this simple action from a motion that you already have in the machine, you can automatically supply oil to many vital parts of your machine by gravity or the cascade system. An easy way to show the operator that the oil supply system is working properly is to provide a transparent dome at a prominent high part of the machine. (Figure 4.14–2)

This simple way of providing for lubrication will be sufficient for all cases where it is just a matter of getting oil to the parts, where speeds and other conditions are such that no oil pressure is necessary. Where oil pressure is required, as, for instance, at the mesh point of a pair of hypoid gears where sliding takes place between the teeth, or at the mesh point of any type of gears where the speeds are high, a pump

4.14 Lubrication

is required that will supply oil at the desired volume and pressure. The oil then must be delivered through thin-wall tubes or small-diameter holes drilled in the gear box housing. It is much better for mass production to drill holes in the casting than to use tubing. These holes are easy to drill, and quite long holes can be drilled in proportion to the diameter when the runout of the drilled hole is not too critical. There are many cases where a tube must be used. Be sure, however, that the tube is properly clamped down and that the direction of the oil stream is correct. Also be sure that the oil has complete circulation and that no pockets or dead ends are trapping the oil.

The purpose of the lubricating system is to make sure that the bearing surfaces are separated by a film of oil; to wash away any foreign matter that may have reached the contact areas of two mating machine parts; and to help carry away heat generated at the contact areas of two mating machine parts.

The lubricating system should, therefore, have a filter of the proper capacity to make sure that impurities picked up by the oil are not returned to the mating machine parts.

Too much oil can be just as bad as not enough. Remember that what you are after is just a film of oil to separate the two mating machine parts. This could be accomplished by immersing the mating parts in oil, but if the speeds are high, the oil would be agitated severely causing churning and subsequent heating of the oil. Air may also be introduced into the oil stream, which would lower the lubricating quality of the oil.

A pressurized oil system can easily be supplied with the proper pump. A pressure switch may be installed in the system which would make contact when the pressure failed and turn on a red light or stop the machine to prevent damage. An audible signal might even be necessary to alert the operator in cases in which production might suffer.

Many types of small machine components, such as small electric motors, are now lubricated for life and need no further attention. In machine tool design, it is considered bad practice to depend on an operator or machine attendant to supply individual parts of the machine with oil. The machine should be so designed that there is only one central oil supply for the entire lubrication of the machine.

Grease packing is all right for bearings and other mating surfaces where speeds are moderate and the grease can be retained. Precaution

Prototype Machine

must be taken so that the grease will not be washed away by oil from other parts of the machine.

4.15 Reappraisal of Prototype Machine

A little time spent in reviewing the project thoroughly before the detail drawings are started is well rewarded by the end results.

As the designer, you must now be entirely satisfied that all design problems have been satisfactorily solved.

Various departments in your own plant and many outside sources supplying components for your machines are now involved, and it is imperative that you have a complete picture of the whole situation.

Since many parts probably are made of castings, requiring patterns, have you worked closely with the pattern shop on costs and ease of manufacture?

Have you worked closely with the process engineer or with the manufacturing departments to make sure that all parts and assemblies are practical to produce?

Have you settled all problems in delivery of purchased parts, and do you have more than one source of supply?

4.16 Process Engineering for Machine Tools

Often, especially where the machine tools are built in large quantities to be sold, the machine tool designer is assisted by a process engineer, who is thoroughly familiar with the equipment and machines available in your manufacturing departments. He should be brought into the picture as early as possible in order to become acquainted with every part and assembly of the machine. There are many things to be considered. Perhaps some parts can only be accurately and economically manufactured by making a fixture. There may also be certain portions of your design that are too weak for accurate machining. You have analyzed your design thoroughly and have found that where the part is used in your machine it is adequately strong and that it would be impossible to make it heavy enough to take the machining pressures. In such case it may be possible to add a false section for support when being machined, a pad that may be easily removed when the part is finished. These things must be known by the process engineer early, so that he can plan the dispatching of the part through the shop.

As your design is progressing, make the layouts as clear as possible without spending unnecessary time. The location and relationship

4.17 Detailing of Prototype

of parts or holes to other parts or holes in the machine may be of great importance to the proper functioning of your machine. If they are, add dimensions and give required tolerances. On the other hand, there may be cases where keys or other parts, ordinarily located to great precision, may not be required to be located accurately on your machine. If this is so, make a note of this so that the process engineer can make plans accordingly, and save time and money in producing the part. Many thoughts probably go through your mind regarding the processing of certain parts. You have come to definite conclusions and you have definite reasons for this. The process engineer does not have the complete overall picture of the machine that you have, so he may make suggestions for changes based only on his knowledge of processing. Some of these changes may be good and may even benefit your design. You are responsible for the finished results so make sure that any changes you make will be beneficial.

In some cases, fabricated parts, e.g., steel parts made up of several pieces welded together, may have to be stress relieved before being finished. Make sure that you can do this in your own plant or that facilities are available nearby for this purpose. You may also have long spindles, shafts or parts in your design that will need straightening before being finished. Do not depend on the process engineer to solve all of your processing problems after the detail drawings are finished. You should have most of this worked out as the design progresses so that the detail draftsman can be properly instructed.

4.17 Detailing of Prototype Machine

The detail draftsman should have a good explanation or briefing of your design before starting the detailing. The time you or your layout men have spent on this briefing is no waste of time, but will pay for itself in final results. There are often cases where possibilities of improvement in design and processing do not show up until you are getting underway with the detailing of the machine. A good detailer, interested in his work, is not just drawing lines and circles, but preparing drawings that will be easy to understand, with just enough views and sections necessary to produce the part. Too many views and sections are confusing. The same is true with dimensioning. Give dimensions on only one view of the same part.

Give all the tolerances desired, but never call for tolerances closer than necessary, and never call for a method of finishing that will give you a closer surface uniformity and finish than necessary, unless it would make no difference in the cost of preparing the part.

Much thought should be given to the dimensioning of castings. Some of the surfaces have a certain amount of draft or taper necessary for drawing the pattern out of the sand mold. If it is necessary to locate surfaces or holes in relation to surfaces having draft, give a starting dimension from a reference point on the surface most important to the proper functioning of the part. Quite often, problems in manufacturing the part to the required tolerances are avoided by doing this.

When preparing a casting drawing make sure that all cores can be properly supported in the sand mold. This can often be done just by adding a hole in a casting wall. Core boxes, the making and placing of the cores in the sand as the mold is prepared, add to the expense of the casting, so if, by designing your casting a little differently you can eliminate some of the core boxes, you have done a good job.

Make sure that there are ways of handling and fastening or clamping the part in the process of manufacturing. Perhaps you have to add lifting holes or provide surfaces for clamping.

4.18 Release for Production

Since the methods of release for production vary with different organizations, and because, after these methods have been established, they become routine procedures within the organization, just a few points of general interest to the designer will be mentioned.

It is good planning to have all parts finished when required, for completion of sub assemblies, major assemblies, and the entire machine.

The production departments play an important part here, so there must be good cooperation between you and the various production departments. The planning starts with you for good end results.

Purchased parts such as bearings, and drive, electrical, hydraulic and pneumatic components should be ordered early enough to be on hand when needed. If forgings are required for some parts, they must also be ordered early.

Castings require patterns, so these patterns may have to be ordered long before the production release so they can be finished when the steel parts go through the shop.

Special parts of fabricated steel also require additional time for stress relieving.

4.19 Important Points to Consider

DATE OF COMPLETION. Early in the design program you probably had a date to aim at for completion of the machine. You are now in a position to predict when the machine can be expected to be finished, and you may have to establish a firm date for completion.

Check up on progress periodically. If this is not done, you may find that parts for your machine have been held up for other work that has had more pressure for production.

Nothing is gained by unreasonable demands, but unless you keep a constant watch on progress all through the design, manufacturing and procurement of purchased parts you cannot expect the desired progress.

APPEARANCE AND FUNCTIONAL QUALITIES. In the follow-up of production, all purchased parts should be checked as they arrive to make sure that they meet the required specifications. Throughout manufacturing and assembly operations, it pays to watch the important parts of your design. If an assembly does not operate as expected, have all the parts removed while you supervise. Even a good mechanic, when under pressure, will once in a while do things ordinarily not expected of him. For instance, a good mechanic was once assembling an important part of a machine for which he lacked a compression spring of the correct length. He assembled two springs making up the correct length. This was an assembly designed for precise repetition. When tried, however, it would not repeat. Removal of all parts was demanded by the designer and the trouble was immediately discovered. There are many parallels of this example, so always look for the unexpected.

TRIALS AND TESTS. Parts and assemblies requiring a period of tests and examinations under actual running conditions should be checked periodically. An accurate report should be given by the department running the test, but periodic checks by the designer are also necessary to get a good overall picture.

4.20 Final Appraisal

After the machine has been tested under actual operating conditions for some time, you will have an opportunity to evaluate performance, based on customer's or operator's point of view as well as on your own observations. You can then arrive at definite conclusions for a future course of action.

CHAPTER 5

Mass Production Machine

5.1 Machine Improvements

In the final appraisal of the prototype machine, you have analyzed production and performance in general. Any desirable improvements, impractical to include in the prototype machine, should now be planned.

There are likely to be other things that do not affect production or performance of the machine which would be very profitable to change, considering cost and other aspects valuable to smooth production schedules.

MATERIAL. In general the material for all parts should be reviewed in the light of mass production, always remembering not to change the material if there is the least doubt that the performance and reliability of the part may be adversely affected by the change. Quality always comes first and in the long run is the best for everybody concerned.

Analyze all parts to determine if it would be advisable to make them from hardenable material. Consider if the surface wear and fatigue of the material would be lessened by the change and if a change would be desirable at all.

STANDARDIZATION. Many parts of a machine can be standardized, such as bearings, bushings, washers, nuts, screws, studs and oil seals. There are, however, also cases where a standard part would not result in good performance, and better results could be obtained by designing the part especially for the machine under consideration. Therefore, analyze each case thoroughly.

APPEARANCE. Good appearance of the entire machine often is a valuable asset.

5.1 Machine Improvements

Properly placed guards will protect both parts and operator, will keep oil or coolant in their proper places and preserve the appearance of the surroundings.

LUBRICATION AND COOLING. Make sure that you have the best arrangement for lubrication and cooling. In many cases tubes can be eliminated by designing channels or drilled holes in the machine parts.

SPACE. Make sure that all serviceable or replaceable parts are accessible and that all components subject to heat generation are placed so that the generated heat can be dissipated as effectively as possible.

5.2 Assembly Drawings

Good records are essential. In building the prototype machine, full size layouts of the various machine assemblies may have been used in the assembly departments. For a mass production machine it is better to have layouts of a small uniform size, each drawing showing one of several views and sections of the various assemblies. These assembly drawings could be made to a reduced size after all detail drawings are finished. Important dimensions should be added to the drawings, and there should be no repetition of dimensions in other views of the same assembly. Drawing numbers of all detail parts should be added to the assembly drawing and placed within small circles of a uniform size, appearing some distance away from the part and connected to the part with a fine line. If it is necessary to repeat numbers in another view, all repetitions should be in dotted circles to make it easier to keep records straight.

Only the most essential instructions, in concise notes, should be printed on these drawings. Since these drawings are of a small uniform size, say 11 × 17 inches, they can easily be photographed in miniature sizes for permanent records, in case the original is destroyed.

5.3 Detail Drawings

Great care should be used by the detail draftsman in preparing working drawings. The ultimate performance and life of the machine depend on accuracy in presenting the necessary requirements to the shop. There should be no sharp corners, especially on hardened parts. Inside corners should be properly relieved with a generous radius in the corner. These conditions are shown in several illustrations in this book. Outside corners are preferably chamfered. In some cases a rounded corner is used.

Mass Production Machine

Distance between holes should be in square moves. If one hole, surface or point is of great importance and the location of other points is important in relation to this, then use this hole, surface or point as a reference. It is bad practice to run dimensions consecutively. Possible errors for each step could be unreasonably high for the last dimension.

When surface finish is of importance, specify the requirements, but never be unreasonable in your demands for this would add to the cost of the part. See Table 4.9–7.

5.4 Photographs

Photographic views should be carefully selected to meet the required needs. Perspective views are, as a rule, the most pleasing to the eye, but in some cases direct views may be necessary to illustrate certain points. Photographs are valuable for record purposes, and useful for the sales, service, manufacturing and engineering departments. They are also of great value to the customer. Pictures taken by the customer of their automatic assembly lines will be helpful to the design engineer when he is contemplating new machines for a certain manufacturing project.

5.5 Operating Instructions

The designer of the machine should formulate a primary draft of operating instructions. This is particularly important for a complicated machine. Close-up photographs of various portions of the machine are also necessary to convey these instructions properly to the operator. It is then often profitable to have the man who runs the initial tests on the machine write the final instructions in easily understood language.

These instructions should first provide all necessary information pertaining to the installation of the machine. Give some general information for lifting the machine with a crane or moving it on skids.

All exterior finished surfaces are usually coated with a rust preventive which must be removed and replaced with oil or grease before operating the machine.

Figure 5.4–1. Straight bevel gear generator
 It should be noticed here that the auxiliary components are placed next to the machine, rather than making them part of the frame, so that the heat and vibrations generated in these components will have no disturbing effect on the accuracy of the machine. These units are then also easier to service. (Courtesy of Gleason Works)

5.4 Photographs

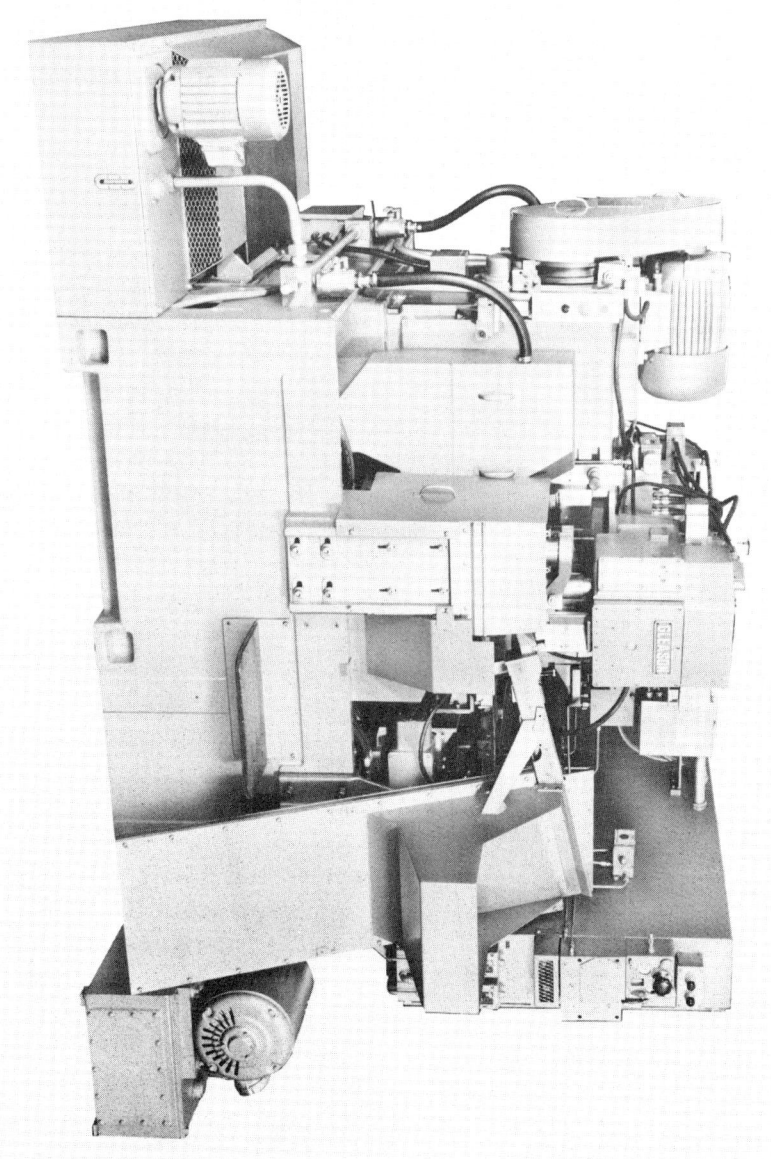

255

Mass Production Machine

Figure 5.4-2. Direct front view of completely automatic hypoid gear finisher with loader, transfer and unloader arrangements for high production of automotive ring gears. (Courtesy of Gleason Works)

Figure 5.4-3. Perspective view of the machine shown in figure 5.4-2. (Courtesy of Gleason Works)

5.4 Photographs

Give proper instructions for connecting to the electric power supply and checking the direction of rotation before operation of the machine. There may also be other power supply sources such as air, for which proper instructions should be given.

The instructions should also contain adequate information for lubrication. The same is true for hydraulic oil, if required. Here it may also be necessary to give instructions for bleeding at certain points before operation of the machine.

Also, if the machine requires coolant, give all necessary information pertaining to use and service.

Finally, give all necessary information about replaceable components and necessary tools.

5.6 Floor Plans

Floor space is at a premium in all manufacturing plants, so in designing a new machine, every effort must be made to make the machine as small as practical and at the same time of sufficient size to do the job.

In planning for floor space, other things must be considered besides the actual size of the machine frame.

1. There must be convenient access to the machine for the material to be processed.
2. There must be adequate room for removing and conveying the processed material.
3. There must be sufficient room for the necessary auxiliary equipment.
4. There must be easy access for service of all replaceable and adjustable components.
5. There must be enough room for operational adjustments and motions.

The simplest way to illustrate these provisions is with a floor plan of the machine, just showing enough lines to convey your information.

Figure 5.4–4. Bardens & Oliver single spindle automatic cutting-off lathe designed to form, groove, chamfer and cut off tubing or bar stock to any desired length, showing limit switches controlling these movements. (Courtesy of R. B. Denison, Inc.)

5.6 Floor Plans

CHAPTER

Reports and Communications

6.1 Engineering Reports

Experience leads to the writing of efficient reports. But until it is acquired, some basic rules can be followed which will help the designer to progress in this important area of his work. If the report is well organized from the beginning, completion becomes an easy task.

If the report is on a machine in the process of being designed and built, reports may be desired at intervals, describing progress, trials and tests, with a final overall report at the conclusion of the design; or just one conclusive report may be desired after the final test of the machine.

In any event, the following should be considered:

1. The reason for the report and its intended objectives.
2. The specific requirements of the report.
3. The main theme of the report.
4. The principal recipients of the report and what they expect of it.

Following are suggested headings to use when organizing the rough draft of the report:

1. Title and Subject. The title must be informative; the subject should be clearly stated, so that the reader knows what the report is about.
2. Abstract. The abstract should be a summary of the principal technical parts of the report which constitute the research work and experiments. It should also briefly state the purpose, significance, methods, and equipment used.
3. Preface. This part of the report introduces the reader to the

6.1 Engineering Reports

scope of the research problems or experiments, and if the project is a major one, it should be divided into sub-headings. A table of contents might be necessary. It should also have a definition of technical terms not generally understood by nontechnical personnel. The reader can then evaluate the report easily.
4. Summary of Observations. Here exact details of all parts of the research problems and experiments are described clearly. This description should be supported with tables, graphs, photographs and drawings, and if unusual methods were used, they should be thoroughly explained.
5. Analysis. If new methods or procedures are deduced, they may have to be supported by calculations, introducing formulas for future use. Do not be dogmatic in your statements unless your findings can be supported by facts. If your research work does not definitely prove a certain procedure, but tends to indicate that such a procedure is desirable, say so plainly and frankly.
6. Conclusions and Recommendations. Definite statements based on facts enumerated in other parts of the report should inform the reader of success, inefficiency or failure. Your recommendations should be definite and fair and may suggest action to be recommended if the project has proved inefficiency or complete failure. Definite statements of proof for the inefficiencies or failure should be made. If no proof was available, you should state what you believe contributed to the failure and give an explanation of why you arrived at this conclusion.

After you have organized the contents of your report, make a rough draft. What you write the final draft of the report, convey your thoughts, observations and decisions as clearly and as thoroughly as possible, leaving no possible question unanswered.

When the final draft is finished, read it for accuracy and check the following points:

APPEARANCE. A report with uniform indentations, good spacing, and clean typing, and with clear illustrations, tables and graphs accurately identified and neatly arranged, is easily read and digested.

CLARITY OF PRESENTATION. In a technical report, many words may be understandable only by men with technical knowledge. Make sure they are defined for the nontechnical reader.

ACCURACY AND CONCEPTION. A technical report is an important document and must be accurate in every detail.

Reports and Communications

CONVINCING PRESENTATION. You must have the conviction yourself that your judgment and analytical methods are correct. Your presentation should then be persuasive, but not forceful.

LENGTH OF REPORT. If possible, just one major project should be covered in the report. If a few sheets will cover all the necessary facts, this should be the length of the report.

TIMING OF PRESENTATION. If the report has been requested by someone else, and a definite time limit been set, plan to meet it on time. If you depend on other departments for some of your research, remember that they may also have other projects underway at the same time, so you cannot sit back and wait. You must check periodically. If you are the originator of the request, time your report to best advantage for good reception. Timing is important.

6.2 Engineering Communications

For successful engineering projects, it is important and necessary to have good engineering communications. Proficient communication methods for promoting an engineering idea or an engineering project begin with you, and complete success depends largely on your attitude.

DIPLOMACY. To gain the confidence of others, you must be willing to listen to their presentations. You must be willing to weigh and consider other ideas. Many ideas, spontaneously presented, may be worthless, but after due consideration may lead you to something, perhaps in the opposite direction. Ideas spring from many sources.

INTEREST. For a successful solution to all problems, it is always desirable to have close cooperation with members within your part of the engineering department, and with members of associate engineering or other departments. This requirement also starts with you. Show an interest in what they are doing, and do not hesitate to offer praise for work well done.

ABOUT THE AUTHOR

OLAF A. JOHNSON, for more than 25 years the senior design engineer of the Gleason Works at Rochester, New York, now heads his own engineering design and consulting firm in that city. Since his graduation in electrical and mechanical engineering from Norway's Skiensfjordens Mekaniske Fagskole in 1920, he has spent his entire career in the United States, working with nationally known firms and the U.S. Navy. He has designed many machines for special purposes and projects, and is the holder of various patents on automatic machines, a field in which he is a specialist.

TJ
1185
.J64
1971

71-847

Johnson, Olaf A.
DESIGN OF MACHINE TOOLS.

DATE DUE

Asheville-Buncombe Technical Institute
LIBRARY
340 Victoria Road
Asheville, North Carolina 28801